WALK THE EARTH
IN OUR SHOES &
PLANT SOME SEEDS
BEHIND YOU

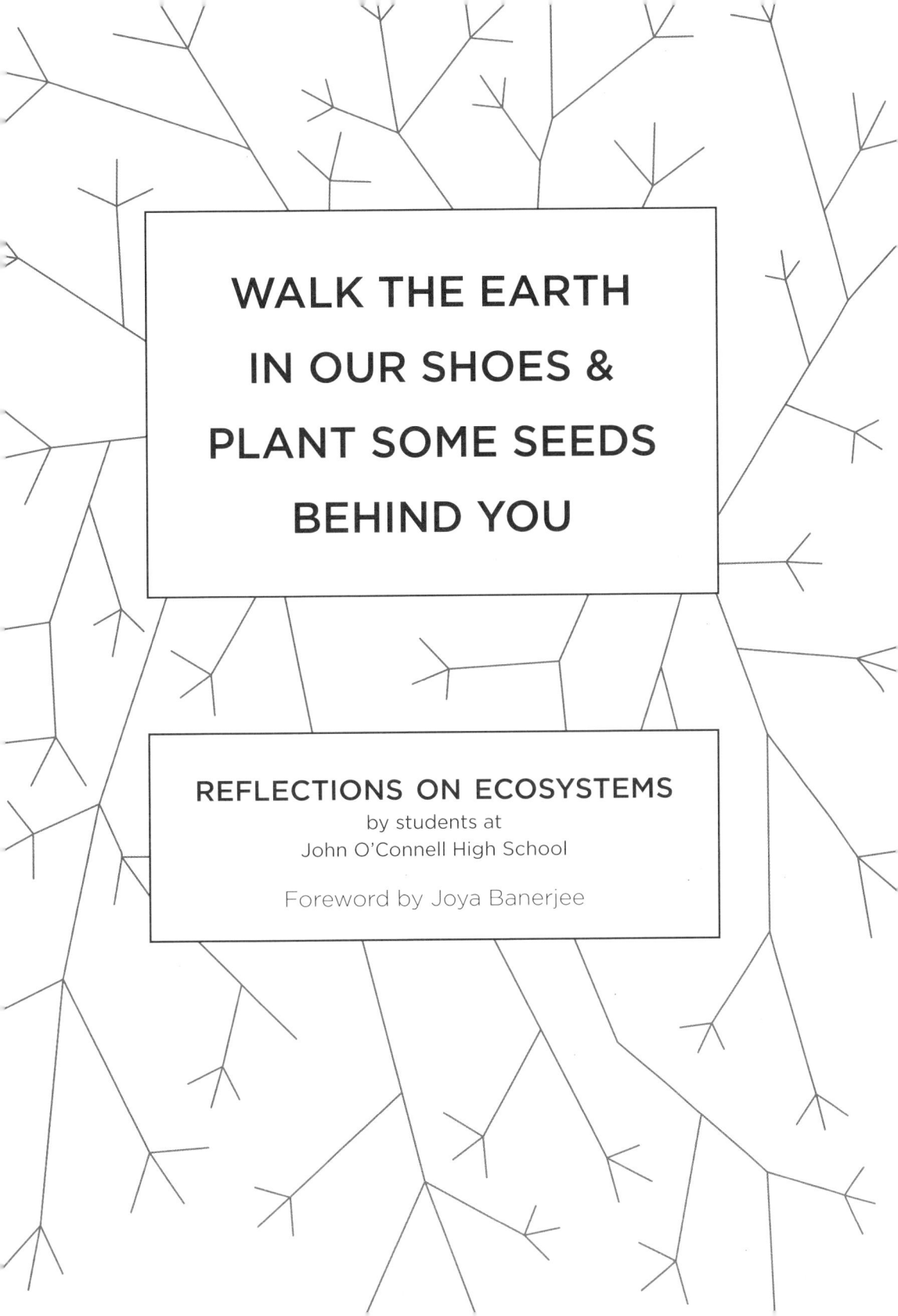

WALK THE EARTH IN OUR SHOES & PLANT SOME SEEDS BEHIND YOU

REFLECTIONS ON ECOSYSTEMS

by students at
John O'Connell High School

Foreword by Joya Banerjee

PUBLISHED MAY 2016 BY 826 VALENCIA
COPYRIGHT © 2016 BY 826 VALENCIA
All rights reserved by 826 Valencia and the authors
ISBN 978-1-934750-66-7

Mission Center
826 Valencia Street
San Francisco, CA 94110

Tenderloin Center
180 Golden Gate Avenue
San Francisco, CA 94102

826valencia.org

PROGRAM MANAGER AND EDITOR
Molly Parent

PROGRAM DIRECTOR
Christina V. Perry

EDITORIAL ASSISTANTS
Dana Belott and Emma Peoples

EDITORIAL BOARD
Elaina Bruna, Tehan Carey, Samantha Gomez, Kaiya Gordon, Gabriela Martinez,
Luna Martinez-Trejo, Izzy Romero-Antoniades, Elijah Romero-Antoniades, Jesus
Savaria, Riley Smith, Amanda Ufeil-Somers, Daniel Spangler, and Dave Struthers

DESIGN DIRECTOR
María Inés Montes

PRODUCTION MANAGER
Amy Popovich

BOOK DESIGNER
Tracy Liu

COVER DESIGNER
Britta Fithian-Zurn

ADDITIONAL DESIGN SUPPORT
Einat Gilboa

COPY EDITOR
Helaine Lasky Schweitzer

Printed in Canada by The Prolific Group.

The views expressed in this book are those of the authors and the authors'
imaginations. We support student publishing and are thrilled you picked
up this book!

Contents

FLYING TOGETHER
What makes a community strong

MY CULTURE IS NOT YOUR COSTUME
Change, challenge, and holding on to roots

IF WE COULD INTERVIEW A WHALE
How we impact the world around us

IMAGINE HOW BEAUTIFUL
THE WORLD WOULD BE
Learning, growing, and taking responsibility

JUST LIKE THE BEAUTIFUL, COLORFUL, DIVERSE SEA
A guide for educators

ACKNOWLEDGEMENTS

826 VALENCIA
Who we are and what we do

EDITORS' NOTE

We've compiled the sources used to inform these essays online for your reference. For works cited and other helpful notes about how to use and read this book, please see *826valencia.org/2016yabpsources.*

Foreword

by Joya Banerjee, 826 Valencia Board member

It is a powerful moment when we realize the impact the world has on us, and the impact we have on the world.

For me that moment was in sixth grade. I grew up in Southern California, in a nice, simple house with a green lawn in front and one out back. This is where I would play with my friends, but on a restless afternoon, these lawns weren't big enough for our imaginations, so we would explore along the "wash," one of the concrete channels that moved rainwater through our town to the L.A. River.

I never paid much attention to the L.A. River, except when it was a spectacle on the evening news—the location of a high-speed car chase or the backdrop behind a reporter during a storm. The L.A. River was just infrastructure that blended into the concrete freeways that weave across Los Angeles. But one particularly rainy day, someone explained to me that the "L.A. River" was once, well... a real river. And it blew my mind.

I learned that the L.A. River once ran freely across the Los Angeles basin. It provided a vital water supply to people and nature, but its floods destroyed communities and took lives. In the early twentieth century, after a series of these fatal floods, we tamed the river by encasing it in cement and removing all vegetation. As a result, water no longer runs its natural course through meandering springs and marshland; instead, it rushes through straight channels from the mountains to the beach at up to forty-five miles per hour.

This channelization has had profound impacts on Los Angeles. Together with other efforts to bring water into the region from far away places, like the Owens Valley, the Colorado River, and the Sacramento-San Joaquin River Delta, the flood control project accelerated considerable growth in the region. It also destroyed natural habitat, divided neighborhoods, reduced open space, blighted communities, and degraded water quality. And it had an impact on me—a restless sixth-grader who quickly became fascinated with infrastructure.

In 1977, Joan Didion wrote, "I know as well as the next person that there is considerable transcendent value in a river running wild and undammed, a river running free over granite, but I have also lived beneath such a river when it was running in flood, and gone without showers when it was running dry." Almost forty years later, Didion's recognition of our complex —and for some of us, fascinating—relationship with water still holds true. We have too much water, we have too little water, and we want water for many different purposes. California struggles to manage this complexity, and as a result, rivers run dry, iconic ecosystems are threatened, and entire communities lack safe drinking water.

But our relationship with water can evolve. If you look across the state, you will find heroic efforts to address some of our greatest water challenges. In Los Angeles, communities are reimagining their river and transforming neglected flood infrastructure into a system that can provide parks, trails, jobs, and habitat, all while strengthening community identity and pride. As a result of community leadership, local, state, and federal officials are directing billions of dollars to restoration and economic development projects along the L.A. River. I go to work every day with the goal of helping California transition to a sustainable water system that can benefit both people and nature. And every day, I think of the L.A. River as a powerful example of water's impact on communities, but also of a community's positive impact on water.

In this collection of essays, you will find rich and varied examples of how individuals and communities impact the world around them. These remarkable authors have researched and reflected upon our relationship with the world, in many ways challenging traditional notions of what we think of as ecosystems and biodiversity. You will learn about the effect gentrification has on housing, rent, and education in San Francisco. You will consider the role that diversity plays in our school systems, our neighborhoods, and even in our video games. You will mourn a community destroyed by an earthquake, and you will recall the last Internet meme that made you smile as it went viral. You will face the magnitude of climate change, its relationship with oceans, communities, and water. And you will picture the faces of California's drought. These essays inspire reflection. They inspire action. They inspire scholarship. And they inspire generations of future students who are finding their voice. By publishing these essays, these authors are having a meaningful, positive impact on the world.

826 Valencia is a place where students can improve their writing skills, develop their voice, and find wonder. That is the impact we hope to have on the world. But our community—the students, families, teachers, volunteers, and the diverse, storied, complex, and beautiful neighborhoods of San Francisco—also impacts us on a daily basis and inspires us to do this work.

So without further ado, please walk the earth in these authors' shoes, and plant some seeds behind you. Enjoy.

Joya Banerjee is a Program Officer at the S. D. Bechtel, Jr. Foundation, a private family foundation based in San Francisco, where she oversees a grant portfolio that advances the sustainable management of California's water resources. Prior to joining the foundation, Joya was an attorney at Latham & Watkins and worked for the New York City Mayor's Office. Joya graduated from the University of Southern California with a bachelor of science degree in Business Administration, and she received a juris doctor degree from Columbia Law School. She lives in San Francisco.

Introduction

by Samantha Gomez, Gabriela Martinez, Luna Martinez-Trejo, Izzy Romero-Antoniades, and Elijah Romero-Antoniades

It feels pretty cool to be publishing this book. It's also hard to imagine: what will it actually feel like when we can see and hold it? What if no one reads it? But no matter what, we're excited to be sharing our classmates' voices and perspectives.

These essays were written by ninth and tenth graders at John O'Connell High School. We are all in a class that combines English and biology. Studying both English and biology in one class can be really challenging. Sometimes it feels like they have nothing to do with each other, and it can be hard to figure out what to focus on.

Choosing a topic to write about for this book was difficult because of the limitations we had, like the writing prompts and the requirement that we write about both science and social communities. That was irritating at first—combining those things is hard! But our tutors from 826 Valencia helped us take the topics we chose and write a lot more about them than we had originally thought would be possible. Some of our tutors helped us make our topics more specific, more broad, or easier to research. Some of our tutors helped us come up with new ideas for approaching the essay that took it to new places. Working with them was great—the tutors were fun! They were definitely a bright side of working on this project. They really helped us bridge the gap between the two subjects.

As the Editorial Board for the book, we read these essays and fixed grammar mistakes, talked about the themes, chose a title for the book, and—most important—we ate dinner. It was really interesting to read the essays by our classmates, some of which taught us new things or told us about world events from a first-hand perspective.

A lot of the topics and ideas in this book are directly or indirectly inspired by San Francisco's Mission District, where many of us live. This neighborhood is changing every day. This is a neighborhood where many of our families, many of them Latino, grew up and lived for a long time. The changes are complicated—some are good, some are bad. There's less gang violence now, and the neighborhood has gotten safer, but the changes have also displaced a lot of the longtime residents because rents are increasing. One thing we've learned, through the process of researching and writing this book and from our own firsthand experiences, is that change is inevitable. But we believe that we should use change as a motivator to better ourselves and to work hard for our families and our futures.

People might think that teenagers like us don't pay attention to big issues, or that we don't care about our world. But we do care. We care a lot. And we want readers of this book to know that we are educated, thoughtful, compassionate, intelligent, and above all, that we're full of love for our communities and our world. We want to inform readers about who we are and why we care.

We hope that readers of this book will keep an open, humble mind. We're hoping to open peoples' eyes—to the effects of gentrification, to why they should be cautious about what they consume, and to see the issues in our world and our communities. We hope you respect the authors' perspectives; the issues and stories in this book have personal meaning for all of us.

We also think it's pretty great that other high school students might read this book. We want any teenagers reading this to gain the confidence to defend what they believe in through reading our words.

Fellow students: your words and your intelligence matter. You and your thoughts are valued, whether they're read by no one or read by thousands of people. Our advice to you is to be on the lookout for resources and people who can help you, like your teachers and writing programs. Don't compare your writing to others peoples' writing. Be grateful for your voice.

FLYING TOGETHER

What makes a community strong

Victoria Louie *is a fifteen-year-old San Francisco native. She is a triple-threat athlete (basketball, volleyball, and softball) who studies anatomy in her free time. She hopes to become a physical therapist. She likes to visit tropical islands and is working on a way to get there without using boats or planes.*

SCHOOL SPICE UP!

School is like the ecosystem of the sea. Students, teachers, and staff are like a school of fish; coral reefs are like the school building; the cleanliness of the water reflects the cleanliness of the school; the sunlight and weather of the day affect the school just like they affect the water. The people in a school with different backgrounds are like the many different species of fish that live in the coral reef. All these factors work together, creating the ecosystem of school and life in the water. I believe having diversity in San Francisco public schools, like biodiversity in water, is beneficial to students' academics and communication.

Biodiversity is the variety of life in an ecosystem. It is important for ecosystems to have biodiversity because the more species richness and interaction, the harder it is for the ecosystem to be disrupted. An example of a biodiverse ecosystem would be the sea. The abiotic factors that affect the ecosystem would be the water, air, rocks, and the sun. The biotic factors of the wavy, blue ocean are plants and animals such as sharks, dolphins, jellyfish, algae, and fish. With all these different types of species, it would be hard for an invasive species to disrupt this ecosystem by eating all the native reef fish. We can compare the importance of biodiversity in the sea to diversity in a school.

Just like biodiversity is beneficial in ecosystems, diversity is beneficial in classrooms, too. Studies at Stanford and UC Berkeley have proven that students perform better on tests in a diverse classroom compared to a non-diverse classroom. The studies also showed that the students were less likely to grow up with racial stereotypes if they grew up in diverse classes. Here is a negative effect of a non-diverse school: A study at Stanford showed that black and Latino students are eleven percent less likely to graduate in racially isolated schools—schools in which one racial group forms more than sixty percent of the student population—than in diverse schools. But, students in diverse classrooms can mutually develop an understanding of perspectives of children with different experiences. This will teach the students how to function in a multicultural environment, such as in a workforce, and how to function and talk with people of different cultures.

For more information, I asked students from John O'Connell High School, a racially diverse public school in San Francisco, how they felt about interacting with people of different backgrounds. Eighty-one percent of the students said they do interact with people of different backgrounds, because they enjoy listening to and learning about other people's cultures and traditions. They also said they enjoyed being around others with various experiences because of everyone's uniqueness. However, some students said that interacting with people of various cultures causes some people to start judging each other. Also, some say that points of view may be too contradictory between people of dissimilar backgrounds.

Even though there are a few critiques about diversity in schools, seventy-five percent of the students I surveyed said they like being in a diverse school such as John O'Connell for many reasons. The students said diversity in the school brings people closer together despite physical and cultural differences and inspires others to be open about meeting new people. Students are able to learn about other people's backgrounds and experiences, and it gives them the opportunity to learn and interact more with people of different backgrounds. Although the school is diverse, there are still majorities and minorities of different groups. But, it is very refreshing to see familiar and different faces of various ethnicities every day at school.

If diversity is good for students, why is re-segregation happening in San Francisco public schools? The number of racially-isolated public schools in San Francisco has jumped in the last three years. That means thirty-nine out of approximately 128 schools in San Francisco are racially isolated. That is thirty percent of all the schools in the San Francisco Unified School District!

This is happening mainly because of the choice system which gives parents more say in which schools their child goes to. In the choice system, parents rank the top schools they want their child to go to. After waiting a couple of months, their child is then assigned to a school that may or may not have been one if their choices, which they then have to attend. If the parents are unhappy with the school assignment, they can apply again and again until the end of their child's tenth grade year. The "best" schools are all at their carrying capacity, the maximum population the school can support, while the lower-performing schools' populations are decreasing. The schools with higher population are taking up a lot of resources, like the great white sharks of the sea, while smaller schools are like the tiny goldfish roaming around the coral reefs that are left with little to no resources at all.

Money plays a factor in this as well. Parents with more resources are able to find out information about the "best" schools in the city. If their child doesn't get into one of the "best" schools, parents with more resources have the time and knowledge to keep trying to get their child in. Parents with fewer resources have less flexibility to deal with the system. When they are searching for the "best" schools, parents are disregarding the other children who are trying to get into schools, too. Even if it is the farthest school from their house, families with more income and time are able to drive or take their child to school every day, whereas families with lower income prefer to go to a local school so they don't have to travel as far. One positive effect of re-segregation in schools is that some families are able to communicate with parents and students of the same background because there are fewer cultural barriers, but the children won't be performing as well academically.

Re-segregation is an example of the tragedy of the commons in our school system. When parents only think about their child getting into the "best" school, it disrupts everyone's quality of education. With segregation taking over, all the students are suffering.

I agree with most of my classmates here at John O'Connell High School that diversity benefits students in all aspects, socially and academically. I believe that students learn and communicate better with a group of diverse students. Diversity gives the opportunity to learn and interact more with people of assorted backgrounds, a skill needed in the workforce. I and many others think it is very refreshing to see different faces of various ethnicities in our community.

My experiences with diversity have been beneficial to me. For example, I am working on learning my third language. Through native speakers at my school, I am able to learn quickly and learn more useful Spanish. This is a great example of mutualism because with their excellence at their native language, they're expanding the knowledge of my tertiary language while maintaining their fluency in their native language.

Just like diversity benefits me, it can also benefit many other students in the San Francisco Unified School District. Instead of giving more choices to parents about where their children should go, the school board should take more control over where the students of San Francisco should go to school. This will help diversify all the schools of San Francisco, just like the beautiful, colorful, diverse sea.

Ixzel Sanchez *is fifteen years old and has lived in the Bay Area her whole life. She likes to go shopping with her friends and her mom. Her goal is to have a good career, and she's considering becoming an immigration lawyer.*

A FLOCK OF BIRDS FLYING TOGETHER

What I know about ecosystems is that everything in them has a connection, like a flock of birds flying in sync together. I'm going to tell you about the ecosystem that I live in.

My name is Ixzel Sanchez. I live in the Bay Area. I live in the Mission District in San Francisco, one block down from Mission Street. The Mission neighborhood seems to me like the most popular neighborhood in San Francisco. It is also dangerous. There is a lot of violence. There are a lot of homeless people, people selling drugs, and cops. The neighborhood is like an ecosystem because of all the interactions between people going on in one day: fighting, shootings, people asking for money, people buying food. There's a lot of human drama and teenage drama, too.

These connections all plug in together in related and also in messy ways. The neighborhood is different from an ecosystem because the interconnections and relationships are often hostile and chaotic, whereas in a healthy ecosystem everything works and lives in sync.

Sometimes I feel safe in the streets and sometimes I don't. You never know what is going to happen. Most of the time when I'm walking I am on my phone, so I'm not connected to what is going on around me. My mom worries about me not paying attention. She gets out of work at midnight so she knows what the dead of night is like on the streets. She starts work at noon but she does overtime, which is why she works so long. She tells me to be safe out there, "because I know people around the Mission and they will tell me what you're doing," she tells me. So I have many eyes watching me, keeping an eye on me. She's been here for fifteen years; she is from Guadalajara, Mexico.

Guadalajara, where both my parents are from, is different from San Francisco. Every country has poverty in some way or another; there is no city without people who are struggling to get by. I think the people are nicer in Guadalajara, though. They don't give you attitude. They see you walking down the street and say "hi" or even invite you over for coffee. I go there with my dad a lot, so I know many people there. Guadalajara has many beautiful beaches, and you hear a lot of mariachi on the streets because that is where mariachi music comes from. The food there is better, too, because it is all homemade. It feels to me like it's more like the natural environment there because of the way people interact.

Why is there violence on the streets of San Francisco? Why do people join gangs? I wonder about this living in the Mission. When people are in gangs out on the streets, they are exposing themselves to death by getting killed or by killing innocent people. I think teenagers go into gangs because most of them don't have enough family or community support. Sometimes not having family support can really lead kids into depression or cause anger toward their parents because they are not getting enough love and attention. Reasons for this of course can include living costs, which is the biggest pressure for many parents. The rent in many American cities, especially San Francisco, is getting so high that many parents basically just work to pay the rent and end up not even having money to feed their kids. Everything gets out of balance.

A family is like an ecosystem because everyone is close to one another. There are also many other organisms around the same neuron, with the neuron as a metaphor for the home. Because of that neuron—the home—everyone in the family keeps a connection.

My house feels way more calm and organized than being on the streets. It has rules: my mom's rules. For example, shoes: if my mom sees me walking in with shoes on she will get pissed at me for not taking them off. She says that bad spirits come in on the soles of shoes and in dirt and germs, so it is a rule to keep shoes outside the door. In the house it's all socks. If my socks are dirty she tells me I have to wash them. She says, "If you don't like it here there's a door for you. I don't have anyone living under my roof if they don't want to. Once you leave here, you're gonna see what life is like when you're on your own." My mom has a schedule for each hour. For one hour when I get home I eat, and then I have to clean, and then do homework, or have free time. After that I walk my dog, and then the rest is up to me.

When I walk my dog I go to Bernal Heights and I feel freedom. Bernal Heights has a nice big park to walk in. You can run, walk, jog, or whatever, and nobody will say anything. I feel free there. It is a very calm place. All the way at the top there is a swing, a secret swing, and it has a nice view. It's a popular place for walking dogs. It's a place dogs feel free also and can play, or meet new friends. My dog's name is AJ, and he's a brown poodle.

Sometimes I see parrots flying in San Francisco, and it reminds me of my freedom. I wish I could fly very high like them, because I would see all the interactions below.

Syncere Austin *is fifteen years old and was born and raised in San Francisco. His hobby is dirt biking. For now, it is his only hobby (his mom thinks he should have more.) Syncere hopes to have a house by the time he is twenty and to be successful in life—right now that means getting a GPA of at least 3.0.*

AFFIRMATIVE ACTION, UNEXPECTED DIVERSITY

What would the world be like without diversity? There would be no excitement, no arguments, and no differing opinions. Everyone would be the same; it would be boring. Diversity is important. It allows us to experience and learn new things. This is important because without learning new things you will never get anywhere in any aspects of life— career, personal relationships, or school.

Even though diversity is beneficial, we don't always see it in our community. Sometimes we need to work hard to make our communities more diverse. I believe that having more affirmative action in our community would increase diversity because everyone will not have the same mindset and will be doing things differently. Affirmative action is so important because it makes life more fair and makes our community more diverse.

One example of a diverse community is my job working at the DJ Project that is part of Horizons Unlimited. The DJ Project is a teen after-school program that teaches DJ skills and does job training. The DJ Project is diverse because there are a lot of different people who work there and are different races or from different cultures. On Wednesdays we do homework the whole time. When I need help with my Spanish homework, there is someone there that can help me. Having a diverse community makes your community stronger because there is always going to be someone that another person can turn to, since everyone isn't the same.

My job is like a biodiverse ecosystem. Biodiversity is the existence of many different kinds of plants and animals in an environment. Similarly, my job has people of different backgrounds who can help you with anything you want. My program has students who represent cultures such as Latino, Asian, African American, and Caucasian. People can benefit from a diverse environment because you can build friendships with new people and make new friends. That way you can learn new things from new people every day. These new relationships are examples of mutualism because both of you are benefiting from the situation. Making new friends benefits each other because you won't be afraid to do anything around someone you know, unlike a stranger.

Another example of a diverse ecosystem is living in San Francisco. San Francisco is very diverse because of all the different races and cultures that live together. You know it's diverse because when you are walking down the street you hear many different forms of music and languages, and see a lot of different races. You could be walking down the street and see someone speaking Spanish, and then there will be a different person walking past you speaking Chinese.

According to Google dictionary, "Affirmative action is an action or policy favoring those who tend to suffer from discrimination, especially in relation to employment or education." Affirmative action helps people who are being discriminated against by giving them jobs based on how qualified they are. Affirmative action increases diversity because it brings in more people who wouldn't have been in schools or jobs because of discrimination.

Affirmative action can benefit our community because by having more than just one type of person telling you things, you can learn more. Another benefit of affirmative action is fairness because you will not get judged, you will be treated like you're supposed to. When there are more people, there is more to be said, so you can learn more. When you learn something in different ways, you learn new methods to do it. This is a result of a diverse community formed by affirmative action.

Although there are many known benefits of affirmative action, there can also be downsides. For example, one of the downsides of affirmative action is that it might place kids in schools that are too hard for them, so they can't succeed. According to an article by Dylan Slater in *The New York Times*, this is called "mismatch." "It's the idea that affirmative action can harm those it's supposed to help by placing them at schools in which they fall below the median level of ability and therefore have a tough time," Slater writes.

In conclusion, in a non-diverse community you will not learn as much as you would in a diverse community. Everyone would have the same mind-set if the community wasn't diverse—you would think the same, talk the same, and have the same opinions. On the other hand, by having a diverse community, people would learn a lot more from each other since everyone has different thoughts, experiences, and ways of living. Affirmative action makes situations fair by not judging people and offering people oppor-tunities to be successful. Affirmative action will make a community more diverse. With more people from different backgrounds, you will want to know more about them out of curiosity. By learning more about them, you get friendship and you get questions answered. Having a diverse community is fun because everything is unexpected.

Jennifer Aguilar Diaz *is a bright ninth grader who grew up in San Francisco. She likes to play soccer at Crocker Park with her friends, often taking on the challenging role of goalie. She has a lot of responsibility as the oldest of four siblings. Someday she hopes to be a nurse or a doctor.*

BE WHO YOU WANT TO BE

"Why are you going to join a sport if you know that you suck?"

When I was in middle school, one main thing that I really wanted to do was join a sport. But there was always somebody who put me down, and that was really hard for me. I became shy and had that feeling of not being able to play any sport. I always thought that maybe I was not good enough, or maybe just that boys would laugh. But even though boys may laugh or say that girls are not as good as they are, girls can be as good as boys can be. I think that it is better to have all-girl schools because that way girls can be more engaged in physical activities and focus more on their school work because boys can be distracting.

Schools are ecosystems that are made up of kids and teachers. When they are in class, they communicate and argue. The relationship between kids and teachers is mutualism because kids get help from teachers, and teachers get paid for teaching and helping kids get a good education. The type of relationship between girls and boys can be like predation. Predation means a relationship between a predator and a prey. Girls are sometimes less likely to participate in activities with boys around because boys might laugh at girls or say they can't do anything. This benefits the boys because they are the ones who are laughing while they're hurting the girls. Biodiversity means that there are different types of life in an ecosystem. Biodiversity is advantageous in biological situations, but in schools, gender diversity is not.

Girls communicate differently when they are around boys than they when they are communicating with just girls around. For example, girls can be shy, quiet, and giggly around boys. When two girls are together they are likely to share anything that they might have in common. When boys and girls communicate, they get shy and are less likely to attend an activity. When both genders are together they like to show off more. In a coed school there is more drama, more fighting, and more rumors. Also, when both genders are together, "Girls comb their hair in rearview mirrors / and the boys try to look so hard." (Bruce Springsteen)

What I have learned is that girls love to play sports and that girls are more likely to join physical activities when they are not around boys. When girls are around girls, they are most likely to be more engaged in physical activities. Some other things I have learned are that in all-girl schools, there can be less drama and there are more positive interactions between girls. They are more likely to share personal things because they can understand each other and girls can be less dramatic when they are around the same gender. Also, some advantages that girls have when they are in an all-girl school are that they get to meet new people and learn and experience a new world with no boys around, because boys can be really annoying.

"The advantages of going to an all-girls school is that there are no BOYS to distract you. Sometimes they get annoying and don't let you focus," said Elsy Osorio in an interview. Elsy is a great girl with a lot of talents. She is also a smart student and she is always honest and focused on her studies. She is also a friend that you can trust. You can tell her anything personal because she will always keep it in herself and not tell anybody. I interviewed her because she goes to an all-girl school, and she has experienced the difference between a coed school and an all-girl school.

Not all people think that all-girl schools are better than coed schools. Some people think that coed schools are better than girls' schools because in coed schools, students get to meet boys and girls. But I think it brings more drama than in an all-girl school. It doesn't really matter who you are or what gender you are, you should be proud of yourself and do the things you want to do or want to be in the future. Don't let anyone destroy your dreams. Always keep your head up and never look back, and always remember to smile.

Sami Al-Homedy *grew up in both Yemen and San Francisco. He enjoys spending time with his friends, going the movies, and eating the wonderful food to be found in San Francisco. Anyone who meets Sami is struck by his sincerity and his infectious smile.*

WARMTH TO MY HEART

The Tenderloin neighborhood in San Francisco means a lot to me, because it's been a home to me. It's where I became who I am. I like that people from all around the world want to go there to chill. I like the weather. It's not too hot and it's not too cold. It's always full of life. I've lived in the Tenderloin for five years. According to the 2010 Tenderloin Census, there are 31,565 people living in the Tenderloin. That's 3.9 percent of the city's population. It's the most densely populated neighborhood in San Francisco. The Tenderloin is a social community that works like an ecosystem, because there are a lot of people who work together to make it a community. Living there made me who I am today.

I have different types of friends there because of the different types of cultures there. It's a really diverse place. When you really start to think about it, someone from a different culture can be like a twin to me and someone from my own culture could be like a snake to me.

The Tenderloin makes me feel like I'm at home. It's a feeling that I can't explain. It makes my heart feel warm when I'm in or next to it. The Tenderloin is a colorful place, and there are a lot of mixed people, a lot of rich people, and a lot of really poor people who all live in one place peacefully. You can tell who is who by the way they dress. There are a lot of cultures in the Tenderloin that live beside each other. There are a lot of languages, like Arabic, Chinese, English, Spanish, Hindi, Vietnamese, Japanese, and some African languages, and everyone speaks slang to each other. Churches and mosques keep a lot of people out of trouble. They are close to each other—they're literally two blocks away. There are a lot of stores (like Walgreens), and that means a lot of food. There are different types of food from all over the world in the mall. As soon as you walk in, you can smell chicken, popcorn, bread, steak, and pepper.

The Tenderloin has help for almost everyone. There are a lot of programs, like the Boys & Girls Club, and a lot of transportation like buses and taxis. There are a lot of jobs. It's easy to make friends there. There are a lot of new parks for the kids from everywhere and the weather is great. There are lots of schools and lots of resources. A lot of people leave murals behind them, and a lot of people hang out outside.

When you walk down the street, you see all of these things and you know that it's diverse. There are lots of different people who have lived in the Tenderloin peacefully for a long time, and they all interact in a good way. This makes it like an ecosystem, because an ecosystem is a community that is full of life and interactions between different types of people or animals (biotic) and things (abiotic). All of these things interact with the environment of the Tenderloin to make a single, complex community.

In the Tenderloin you feel a sense of community. There are all the people from schools helping out the kids by taking them to the park, taking them home, and feeding them. I see kids playing with each other. People chat in the stores and leave happy. I see Girl Scouts selling cookies to whoever walks by. People sit down and watch movies together. This is what makes it home to me. This is what a neighborhood looks like when it's balanced.

Now I see new buildings being built, new parks, and new people coming in. Some of the old buildings are getting destroyed. Some people have left the neighborhood. Now I see new faces when I walk down the street. I see more police cars in the neighborhood. Soon everything we live with is going to change. It feels a little less like home than it used to. All these changes are disrupting the balance. I don't know if anyone can do anything about it. But I hope that they can, in a good way.

The Tenderloin is a lot of things. Right now it's changing, but it's still a great place. It feels to me like one of the safest places in San Francisco because it feels safe in the morning.

Izzy Romero-Antoniades *grew up in San Francisco with her brother. Her hobbies include volunteering at Rocket Dog Rescue and the SPCA, and going to jazz dance class. When she graduates she hopes to go to Cornell University, major in animal sciences, and become a veterinarian.*

ROMAINE CALM, LETTUCE SHOP ORGANIC!

My family owns and operates a ranch in San Antonio and Lemitar, New Mexico that has a variety of livestock, including a baby alpaca. On the ranch they use their stock of cows, chickens, and pigs for meat production. I started visiting their ranch when I was three years old; I would ride the horses, take care of the chickens, and take care of the baby alpaca. During my time on my uncle's ranch in New Mexico, I observed how well they treated their animals. They fed and watered them, and let them roam free as well. There was no harsh treatment of the animals and the animals' welfare was a top priority at the ranch. Ever since that experience, I viewed the livestock more as companions or a friend rather than just being there for profit. I was able to be affectionate toward the animals as well as help out with the ranch by making sure that the animals were well taken care of. Even though they raised some of their animals to be slaughtered, it was still a main priority to make sure the animals were well cared for.

These experiences have shaped who I am, and it helped me realize that I want to make a difference in animals' lives and animal welfare. I want to bring attention to animals' lives and show people that their lives matter. I want to be able to have a voice in the community and I hope to bring others with me. What I hope to achieve with this essay is to give the reader an understanding and awareness that proper treatment toward farm animals should be a priority, if not a top priority, to buyers, ranchers, and farmers because it leads to better health for consumers as well as the environment, and because it improves animal welfare.

Farming is like an ecosystem because it contains a community that includes ranchers and farmers, the animals, or product in this case, and consumers, or anyone who consumes meat. It functions like an ecosystem because the farming community contains abiotic and biotic factors including land, as well as the animals, ranchers, and farmers. These factors interact daily, which make them an ecosystem. A typical day on a farm starts like this: the farmer or rancher gets up at about 5 a.m. He or she shoves down a hearty breakfast and gets on the job. To farmers, there's no such thing as a weekend. It's a 24/7 type of commitment. There are a lot of commitments a farmer has toward their animals. You would think they would want to be compassionate and care about their well-being. Well, some current farming practices prove otherwise.

Sadly, industrial farming practices often lead to psychological and physical distress for the animals due to their being restricted to small spaces and unable to roam. Along with this, the animals are consistently fed unhealthy diets and injected with hormones in order to increase their meat production. This shows that factory farming is not only cruel to animals, but puts the animals' and customers' welfare at risk in order to make a profit. The animals not only suffer from physical and psychological abuse, but customers can suffer due to the antibiotics given to animals because of the poor living conditions and the hormones given to animals to increase meat production.

On the other hand, organic farming is different. "Organic farming refers to agricultural production systems that do not use genetically modified seed, synthetic pesticides, or fertilizers. Some of the essential characteristics of organic systems include design and implementation of an organic system plan that describes the practices used in producing crops and livestock products; a detailed recordkeeping system that tracks all products from the field to point of sale; and maintenance of buffer zones to prevent inadvertent contamination by synthetic farm chemicals from adjacent conventional field," according to the Organic Farming Research Foundation (*ofrf.org*). Many people ask if organic farming is more humane than conventional farming. Well, the answer is yes! Organic farming is more humane, because the USDA mandates specific living conditions for organic livestock, including access to pasture, access to shade and indoor shelter, and an exercise area. Living conditions must be appropriate for livestock based on stage of life, the climate, and the environment.

Many customers who shop organic have very high expectations for animal welfare due to the amount of money they're spending. I conducted a survey for Whole Foods and Safeway butchers and customers. I asked them a few questions regarding what goes into their meat purchase. Is it the quality? The price? The brand? What do you think is personally better? The majority of the people who were surveyed said it was the quality as well as the price. See, if you shop at Whole Foods you can always get certified organic. If you shop at Safeway or, say, Foods Co., you're not guaranteed to be purchasing organic meats because of the variety of suppliers. Certified organic products provide customer satisfaction due to their great taste, no GMOs, no fertilizers, no pesticides, and no chemicals!

California has the biggest organic sales per year. We make over $2.2 billion each year selling organic products throughout the state. According to the 2014 Organic Production Survey released by the USDA's National Agriculture Statistics Service, in 2014 there were 14,093 organic farms producing on 3.7 million acres. Ten states accounted for seventy-eight percent of sales. California alone accounted for forty-one percent of all organic sales and led the way with 2,805 organic farms. Go Cali!

Why is organic farming so expensive? Well in general, organic food costs more than conventional food because of the laborious and time-intensive systems used by the typically smaller organic farms. You may find that the benefits of organic agriculture offset this additional cost. At the same time, there are ways to purchase organic while sticking to your budget. Consider the following when questioning the price of organic: the price of organic food reflects the true cost of growing. The price of conventional food does not reflect the cost of environmental cleanups that we pay for through our tax dollars. Organic farming is more labor and management intensive. But remember, when you shop organic you are supporting farmers' careers and animal welfare.

In light of the information and research here, I believe we should support organic farming as well as organic and humane farming foundations such as the Organic Farming Research Foundation and Simply Organic. Want to help support your local farmers? Go to farmers markets, buy organic foods, or donate to a local foundation! This is important because you are not only supporting farmers, but you are supporting humane treatment toward animals and their welfare. I hope that my essay has inspired you to help your community. You have the power to make a difference in farm animals' lives. Start now!

Alex Hyman *is originally from the cliffs of San Francisco's Richmond District. He now attends John O'Connell High School where he spends his sophomore year passionately pursuing his studies in paleontology and biology, as well as creating and exploring fantasy worlds of role-playing and video games.*

I HAZ MEMES

What if I told you… memes are complicated? Likely, you'd have no idea what I'm talking about, right? Whether you know what "meme" means or not, you have very likely at least heard it somewhere. Memes hold more meaning to the world than it may at first appear. You can throw the most simple of things out into the Internet, come back just one day later, and you will have something you can't even recognize. Memes are like living things in the ecosystem of the Internet because they have interactions with people, allowing them to gain a longer life span.

Okay. First question: Why? Probably the most fundamental question to anything, you might be asking, "Why the heck am I reading this? In what way is this important to me in the least? Why should I care?" Well, if any of these questions can be applied to you then let me take a few moments of your time to shed some light on them. For one, memes are everywhere. You almost literally can't go anywhere on the Internet (especially media sites like Facebook or Tumblr) without finding some kind of meme. As for the science, it's actually quite interesting and surprised me as I was researching it, so I hope to share that with you.

Although memes today are considered a joke or funny thing, there was a legitimate definition that was created. An evolutionary biologist, Richard

Dawkins, came up with the word "meme" and defined it as "an idea, be-havior, or style that spreads from person to person within a culture" in *The Selfish Gene*, a book about evolution and natural selection.

The concept of memes has gone viral to become the much better-known Internet memes. These are what most people think of when they hear the word "meme." An easy example of an Internet meme would be the "Grumpy Cat" meme. For those who aren't familiar with this trend, which I'm sure is a reasonable amount of people, this meme was first created in 2012. It appeared as a picture on Reddit of a cat with feline dwarfism and an under-bite, which gave it a grumpy-looking appearance. It is worth mentioning that most popular memes consist of either text, an image, or both. Here's an example of a meme with Grumpy Cat as an image with the following text: "Good mornings? No such thing." Grumpy Cat exploded in popularity and quickly spread across the Internet, almost instantaneously becoming an Internet meme. Although older memes like this can become overshadowed by newer ones, all memes can come back if their popularity holds. The Grumpy Cat meme is a very stable member of the Internet ecosystem.

Just like living things need resources to survive, memes "survive" on attention: "The moar they git the moar they're gud, scrub." Okay, now that right there would be an example of what's called *lolcat*. Lolcat is an altered language that is quite common with image memes. As you can probably tell, memes enter and exit the Internet in different forms, from images to language, from historical events to pop culture references. These are the tools they utilize to literally catch people's attention, which is the food they need to survive.

The way I see it, the relationship between memes and people is what is known as mutualism, where both sides benefit. In this case, the memes benefit because they get their need for attention satisfied, and the people benefit because they get enjoyment or a laugh, just something to brighten their day. Memes are very lifelike, using the Internet as a sort of habitat. Calling the Internet a habitat and a meme being considered alive are very strange ideas. However, what determines whether something is alive or not? These memes behave a lot like living things if you look at them carefully. There are different species, the Internet is like their home, and they are all trying to survive.

Memes as a living thing—pretty tricky in terms of classification! Would the different types of memes be considered a species? Would a group of memes with similar styles be considered its own family? Would each individual altered image be a member of said family or species? These questions bring me to my next point.

An interesting feature of memes is that they have become so adaptable that they are able to sort of mix and match with each other, becoming a sort of hybrid. This ability in particular is rather odd in terms of a competitive nature. Almost all living things are naturally competitive, even humans, which is why this is so strange. At first glance, they do seem to be competing for attention. Yes, if you look closer, it's more like they coexist with one another. If these memes are truly living things and still act this way, perhaps they have evolved in a way that lets them work together for the same resource instead of fighting over it.

Keila Mejia *is from San Francisco, California, and is fourteen years old. Keila enjoys playing sports, especially wrestling and volleyball (she also plays basketball and soccer.) Keila wants to go to college after high school, and looks forward to visiting Los Angeles soon.*

WHO WINS THE MEET?

Keila on me, Keila on 3, 1-2-3, KEILA! Hearing my own name makes me feel not scared, but tense. I'm hungry enough, but I don't know who I'm going up against. Practice has helped me enough to know that my hunger will be pleased after the whistle blows.

The wrestling team at John O'Connell High School is an example of a non-biodiverse ecosystem, because there's not a lot of variation on the team. Making our wrestling team a biodiverse ecosystem would consist of having a spread out variation of weight, gender, and ability of the wrestlers.

Biodiversity is the variety of species in a habitat, which is life in an ecosystem. Biodiversity would benefit the wrestling team. For example, on days of our meets when we go against schools that have a really big, diverse team or even a small one, our struggle against these teams is that they have more than one person in a weight class and can put someone with two to four years of experience against a first-year wrestler, just to get that win.

The interactions on the team suffer if there isn't biodiversity. The team could have a much stronger base if we had a more diverse team.

For example, predation is when a stronger organism feeds on the weak. This happens during meets too. It's not just physical strength, but mentally whoever is stronger will have the control. It also depends on how you have practiced and whether you have practiced the right way. If not, then the other wrestler has control of you.

Imagine there are two wrestlers at wrestling at 120 pounds, but wrestler A has been cutting weight since last season, working out, and practicing. Meanwhile, wrestler B is new at wrestling. When they both step on the mat, wrestler A will get a pin on wrestler B because he's taking advantage of wrestler B's condition. This can also be called predation, which is when a stronger organism (predator) feeds on weaker organism (prey).

Competition also exists in this ecosystem. Competition consists of two organisms who want the same limited resources. As wrestlers, our limited resource is getting that win; it's something only one of us can have and the other organism can be harmed in the process.

I interviewed our wrestling coach here at John O'Connell, Bob Gamino, and asked him how many girls he has coached since he's been here. His answer was that he has coached seven girls since 2003. Bob also said that we used to have fourteen or fifteen wrestlers and eleven weight categories, and now our wrestling team has less than ten wrestlers and barely seven weight categories. This information shows us that there is a lack of biodiversity on our wrestling team.

Our wrestling coach shows altruism in the way he spends his time to come in and coach us and practice with us, when he could be at home taking care of his baby girl. Altruism is when an organism acts in a way that benefits another. My wrestling team has definitely had examples of competition and altruism, but we have struggled more with having a diverse team.

You could see a big difference between our team and Washington's wrestling team in our first match. When we walked into the gym, we could see that their team was bigger, which meant that they probably had more than one person in a single weight class. It was a team with more than one female wrestler, and they also had more experienced wrestlers. I honestly want to have a diverse team: a team where we have more than one person in a weight class, have more than seven weight classes, have more skill, have more than one female on the team, and have a variation of races on the team. I want it to be open to anyone.

Making our wrestling team a diverse ecosystem will consist of having a spread out variation of weight, sex, ability, and types of wrestlers. As you can see, diversity could be important to any team.

Demyiah Simpson *is a San Francisco native. He lives in the Excelsior District with his father and grandfather, and is a big brother to his two younger sisters (who sometimes annoy him to no end). Demyiah likes to play basketball, listen to hip-hop music, and eat tacos. He plans to go to a university and become a dentist. Ouch.*

USING TECHNOLOGY TO COMBAT CLIMATE CHANGE

Did you know that every year, the earth is hotter than the year before? Trees are dying, there are multi-year droughts across the United States, and terrifying extreme weather across the globe. Can we stop this? I say yes, as long as we all adapt to new technologies and conservation efforts.

California grows more than eighty percent of the country's fruits and vegetables, in addition to a huge amount that is shipped to foreign markets. Eco-friendly and user-friendly technologies are very important because they are our weapons for combating climate change. The utility companies provide lowered costs or rebates on appliances that we can use in our daily lives to conserve water or electricity, which is a big step. Over time, all the conserving will start to add up.

Rebates are a wonderful incentive because people are rewarded with money for buying eco-friendly appliances. Since 2010 we have been able to get tax rebates for some electric cars. Electric cars are so important because they release less carbon, a key contributor to climate change.

A lot of the technology is geared toward the agricultural industry. From drip technology to tractors with GPS that measure the exact amount of water needed in plants anywhere in the field, technology like this helps farmers save water. Farmers can waste a lot of water on plants trying to water them. This new technology helps keep the vegetables and fruit prices affordable at the produce market, and if we can conserve water like this on a daily basis, we would have extra water to use.

If everyone plays a part, we can achieve mutualism in our ecosystems. Mutualism is when two forces help each other positively and they don't harm each other. So if we help the earth by conserving water and not polluting the air, we can get the resources we need and help combat climate change.

It's critical to have a smart water sense. Water is a finite resource and is so important to humanity and its development. Can you imagine being at home, going to your faucet to turn it on, and nothing coming out? That's one of the scariest things to me, and I'm just thinking about it. Imagine if it was a reality. I never want to experience something like that in my life, and I don't want you to either.

Maria Moreno *was born in San Francisco. She plays the flute, viola and guitar, loves to paint, and is a very creative person. She would like to use her creative thinking to do something with animals. Her future goals are to travel and to spend time with friends and family.*

ADOPT A TIGER TODAY!

Do you have an animal that you love and want at your side? What if this animal suddenly disappeared? How would you feel? I would feel horrible, devastated, and would do anything to see the animal again! In Australia's ecosystem, biotic factors like animals are disappearing at a tremendous rate. These animals include koalas, kangaroos, red pandas, the caracal cat, and many more. These are just some of the animals that are being affected in Australia. According to *The Guardian*, "The rate of [biodiversity] decline in Australia is up there among the worst in the world." Many animals are losing their homes and the ability to survive because of agriculture, land being cleared for housing and urban development, the effects of grazing, climate change, and much more. But we could change this by helping save the animals' homes, and this is what matters to me. This is important because I love animals, and seeing them disappear is terrible and heartbreaking! I chose to write about this topic because many people care about the animals as I do, and if many people care, then maybe the remaining animals can be protected.

Australia is recognized around the world for its large amount of untouched wilderness. On the other hand, for people who have recently been there, it is clear that Australia's pristine wilderness is becoming a myth quickly. Evidence of human impact is everywhere. Australia is a small, unique country. It contains more than five percent of the world's plants and animals; eighty-seven percent of them are found nowhere else on earth! Also, Australia, that little continent, has seventy-five major terrestrial ecosystems, and each one of these is composed of hundreds of smaller communities of plants and animals. This shows that Australia, although it's a small land mass, has extraordinary biodiversity. The term biodiversity means the variety of life in an ecosystem, or the different kinds of plants and animals in an environment.

Although this continent has seen lots of losses, it still has many beautiful places like the city of Broome, which is located in the Kimberley region of Western Australia. This city is a quiet place where you can see the blue, clear ocean. Also there's Kangaroo Island, an island that is full of natural landscapes and untouched bushland. It's a natural sanctuary for wildlife and birds and is surrounded by water. Although it has its challenges, Australia has many beautiful and unique places.

My concern is that what is happening to the special animals in Australia is the ecosystem that they depend on has suffered huge losses and many species have been devastated. The causes of this are numerous. Some of these causes are agricultural, clearing land for housing and urban development, the effects of grazing, and climate change. Climate change alone has a major impact on Australia's plants, animals, and ecosystems, and that presents significant challenges to the conservation of Australia's biodiversity. Clearing land for housing or for urban development is one of the major ways that people cause animals to become endangered. When people clear the area, to make factories for example, they are eliminating animals' habitats. This relationship is called competition, which is when two organisms fight and they both get hurt or harmed. When you clear to make buildings and factories, you're competing with animals for space and this affects us and the animals.

I believe that people want to know how can they help and what they can do to save endangered animals. In fact, some people have protected and restored habitats. For example, at Instituto Terra a man called Sebastião Salgado, a photographer from Brazil, purchased a large amount of land, Bulcão farm. In 1998 there were only foxtail plants there and you couldn't see anything. There were no plants, no water, and no animals. Ten years later, this man and his wife were determined to restore the habitat by planting hundreds of thousands of trees around it. As a result, thirty species of mammals have come back, along with 168 species of birds, and fifteen species of reptiles. This is huge. WOW! High fives to this guy and his wife! This forest has a potential for a great impact on Brazil's ecosystem. This project is an example of the relationship called mutualism: a type of relationship in which both members benefit, so it's a "win-win." This farm is a win for animals because now the animals on this land have come back and they have a home. It's a win for humans because many people come to that forest, study it, and learn to make better choices. Also, the trees produce more oxygen so we can breath cleaner and better air. This is really impressive to me. Although it's not easy to plant that many trees, if we all work together, this dream of protecting the habitat could come true. Everything is possible if we try.

One of the ways I can help is by sharing how serious this issue is. I know that we could protect animal habitats by not cutting down trees and not killing animals. I would like to see and experience more conservation models we could imitate. I will try my best to help the animals and to be one of these models.

Endangered animals are becoming more endangered than ever. Will you let that happen? Will you help? Take a step. Have a voice and help the animals no matter what you do! If you help the animals you will feel proud of yourself. You can even "adopt a tiger" through the World Wide Fund for Nature, like my tutor did.

Christley Griffin
*was born in Oakland and grew up in San Francisco. He is fifteen years old and likes drawing. He also loves to play video games (*Super Smash Bros. *in particular). With his favorite character, Bowser, he can shoot fireballs against his enemies. One day he hopes to make his own video game, featuring robots alongside magic.*

UNICORNS (YES, THERE ARE UNICORNS)

One of my favorite games is *Super Smash Bros. 4*. I like this game because it is fun. It's unique from other fighting games because there are many different gimmicks and characters. Diversity of characters in video games is a good thing because a diverse roster of characters is more interesting for the player. It is easier to identify them, express yourself with the use of different strategies, and understand the characters and the mechanics.

"Biodiversity" means different types of life in an ecosystem. Biodiversity helps an ecosystem because different organisms serve different purposes. Diversity is also good in video games because if everyone were the same character it would not be as fun as if everyone were someone different. With diverse characters, games could be more fun. According to one of my sources, someone who posted on a video game message board about why they think diversity is good stated, "Every time I play a game like this, I try my best to make a balanced team." A diverse team can take care of each other's weaknesses.

In a video game I would make (which might be a fighting game), there would be a lot of different characters. Many of the characters would be based on mythical creatures, reptiles, dinosaurs, or even other characters from other games. In my game I want to have characters that you could easily identify. One of my potential characters is called "Dragura." She can turn into a dragon (not a very big one, sadly). She has magenta skin with purple spikes, and she fights evil. My video game world would be similar to ours because it is basically our world, but everyone looks different. Unicorns (yes, there are unicorns) can use their magic to get things out of reach from other characters. This is an example of altruism, because the unicorns would be using their energy to help others. The characters have different powers and abilities.

An example of mutualism in my game is between Dino and his friends. In science, mutualism is a type of relationship in which both members benefit. In my game, Dino's friends would help each other. Dino is a dinosaur. He's yellow with an orange underbelly. He carries a red block on his back (for some reason). He has hands like old cartoon characters and he has a top hat. Dino's friends help him fight in battle against opponents. If Dino didn't have his friends to help him, he would probably struggle, and he wouldn't be a good character. In my game, I want everyone to be equally powerful. No one is the best character, and Dino wouldn't be so powerful without his friends.

In *Pokémon*, the player and the player's rival relationship is an example of competition because both of them are trying to be the very best and to see who's stronger. But both can't be the best—there can only be one. *Pokémon* is an RPG (role playing game) where each player can have a maximum of six Pokémon to fight for them. This is like competition in a natural ecosystem because when the players compete, only one of them will win.

An example of predation in my game is between the character Demimorph and everyone else (that is playable) in the game. In science, predation is a type of relationship in which one organism benefits while the other organism is harmed. Demimorph is a sort of princess. She has red hair. She has the ability to morph any body part she wants or morph herself entirely into her true form in a way. In Demimorph's story, or arcade mode, she will have to fight every character in the game. At the end of it, she absorbs the others into her body like a predator.

I don't even remember the first time I played video games. I love video games like Donkey Kong loves bananas. Diversity of characters in video games is good, just like biodiversity in ecosystems. This is because they display mutualism, competition, alturism, and predation. Without these things, video games wouldn't be as fun.

Cortney Shelton *was born and raised in San Francisco, California. After school she works with preschoolers in Bayview-Hunters Point. She likes to sing Mariah Carey in the shower and suffers from the giggles in class. Cortney hopes to make her family proud by becoming a pediatrician.*

WALKING A MILE

Try this: put yourself in these shoes. Imagine you're ten and this is your first day in a new class and you're the only one of your race. Imagine you and your family are having a party and your neighbor calls the police on you. Imagine you and your family just moved into a new neighborhood and you're the only family of your race. What would you do?

You know how they say you can feel like somebody if you walk a mile in their shoes? But there's one thing: their feet could be bigger than yours, or smaller. If the shoes you're walking in are too big then you're going to look like a clown, and if they're too small then they will hurt your feet. If two people were to switch their lives, both would feel like they don't belong or would feel like they're missing something. The question is, how long would it take for them to adapt to those shoes? Because what I'm trying to say is that people adapt to different social situations at different speeds, just like animals adapt to changes in ecosystems.

When my family and I move, my sister adapts slowly like a fragile animal, like a baby bird. She is very shy and doesn't really like to meet new people, which is why she will adapt slower. When we first went to a new school, her grades dropped. She was very shy and didn't want to interact with people. She kept to herself because she wanted to fit in and know people liked her. She was away from the neighborhood that she knew and this affected her life. She spent her energy trying to change herself so people would like her. Instead of trying to change her grades, she was thinking about whether people liked her or not. When she finally did adapt, her grades improved. This is similar to a baby bird that learns how to deal with life after its tree gets cut down. Just like my sister, the baby bird has to spend energy finding somewhere else to spread its wings and adapt to a new tree or a new forest and adapt to the other animals around him.

My mom is more like an ant, because she adapts to different communities like an ant adapts to different ecosystems. She invites new people over to her house right after she settles in to her house, which takes two days at the most. She's very social, and she can talk all day if you let her. She's not sensitive about other people's opinions. She also came in the classroom on the first day of school with me and my sister to make sure everything was to her standards. She wanted to make sure we were able to get an education right away. I would compare my mom to an ant because if there is a fire in the forest, the ants will go underground and wait for the fire to go away, and then come back out and settle in as if nothing ever happened. In other words, ants adapt fast to different ecosystems.

I would compare my auntie to a squirrel, only because squirrels adapt not slowly and not fast. It takes time, but not too long. When she moved, she invited the neighbor's kids over to play. That helped her kids get friends and helped her get to know the parents. My aunt went through the children to get to the parents because she doesn't want to rush the friendship, unlike my mother who went straight to the parents. She didn't go to the school with her kids on the first day, but she did go and put both of her kids in the same class after the kids told her that they didn't feel comfortable being split up. This shows that she's not fast and not slow at adapting; she took everything one step at a time. This is like a squirrel adapting to a change because if somebody cut down its tree, the squirrel would have to leave its original home and see if it had hidden nuts in the trees. That shows that squirrels don't adapt fast or slow either. They have a backup plan, but then again it's kind of a slow process for the squirrels because they have to think about and spend most of their energy on how much they can carry to their new habitat.

The difference between human communities and animal communities is that people have feelings. I'm not saying that animals don't have feelings, but they're different. If we were to put an animal in a community where the animals all look different, it wouldn't really feel like it doesn't belong. But when we put a human in a different community, like for example if a white person moves into a black neighborhood, he might feel unwanted because there is no one there like him.

Everybody adapts to different situations at different speeds because everybody is different in their personalities, emotions, and how much experience they have with moving or adapting to change. Some people like to move step by step, some people like to move slower, and other people like to just go in and do it. Animals adapt to different situations differently also. Decomposers like ants can eat a variety of things, and consumers, like squirrels, have to wait for the right season for the nuts to drop from trees. This shows that whether you are an animal or a human, we all adapt at different speeds.

Sierra Rivera *is a ninth-grader at John O'Connell High School. She's particularly interested in history and anything related to "the olden days." Writing this essay helped Sierra understand the Holocaust, a chapter of history about which she is especially curious. Playing sports makes Sierra happy (basketball and softball are her favorites.) She comes from a large family—she's one of six siblings.*

THE HOLOCAUST AS AN ECOSYSTEM

"At every step, somebody fell down and ceased to suffer."
                                        - *Elie Wiesel,* Night

According to the United States Holocaust Memorial Museum, there were nine million Jews in Europe, but after the Holocaust there were only three to four million left. Before the Holocaust, the Jews in Europe had a normal lifestyle, great jobs, and even were happy. But once Hitler came along, nothing was the same.

The Jewish community during the Holocaust was like an ecosystem because in an ecosystem, if one thing gets altered, it messes up the whole cycle of life. In the Holocaust, once Hitler came to power, it changed the life cycle of the Jews. The Jews were tragically affected during the Holocaust and this had a huge impact on the lifestyle and social community of the Jews in Europe.

Most of the nine million Jews in Europe at that time lived in Eastern Europe. The Jewish people lived in towns and villages called *shtetls*. The shtetls were each like their own ecosystem because they held many biotic factors, like the people, and they also had abiotic factors. They had non-living things like clothes and furniture. Their children attended school regularly, but some left early to go work in a craft or trade if their family was poor. The children had many opportunities to interact with the biotic and abiotic factors around them. Each person had his or her own niche; each had a role to play, which maintained an ecosystem with a healthy equilibrium. The Jewish people gave to others in the community, even though some didn't have a lot to give. Even people who were really poor often helped the community.

Once Hitler came to power there was a huge dramatic change in the Jewish community. Families were split apart, people who did not cooperate were killed, and some hid away in the shadows. When Hitler came to power, he scared the people of the Jewish ecosystem, saying that they were not a part of his guidelines for being German. Hitler was the predator in this situation. He wanted to erase the Jewish ecosystem. The Jewish people were his prey.

Concentration camps were places for Jewish people to be held. The Jews were transported in trains, sometimes thousands in one train. They had little water and no place to sit down. Sometimes it was even difficult to breathe. When they arrived at concentration camps and ghettos, they all had a tattoo of a number to show who they were. They weren't allowed to look different, so the Nazi soldiers shaved their heads bald and dressed them in the same clothing. All these Jews were taken and transported to concentration camps, and the old people and children were taken to the gas chambers. Only children and old people were brought into the gas chambers because they were not able to work. If people tried to escape, they were shot and killed in front of everyone.

When the Jews left the camps, many of them fled to different countries or went to find their families. Many of the Jews were traumatized. The social community was never the same again because many of the Jews had been killed off and many of them immigrated. Their home was important to them, but it can be hard to return to a place that caused you so much pain. Don't you agree?

In conclusion, the Jewish ecosystem changed dramatically because of how Hitler treated the Jews and how they had to change their lifestyle. We haven't had another Holocaust situation, but to compare it to modern day, we have many communities being forced from where they are and many stereotypes about groups that are different from us, like the stereotype that Mexicans are coming to the United States and taking our jobs, or that all Muslims are only here to bomb us. This kind of thinking should remind us to come back to what happened during the Holocaust, and make sure we prevent another one.

Patrick Garcia *was born in Manila, and lived there until he and his little brother came here with his family five years ago. In eighth grade he was the flag football champion, and now he's on the varsity basketball team. He dreams of attending business school, creating his own tech business, and making millions.*

CALIFORNIANS TRYING TO SAVE WATER

It is a habit for a lot of people to leave their faucet on while brushing their teeth. In a Colgate commercial that aired during the Superbowl, it shows that if we leave our faucet on while brushing our teeth we will waste gallons of water. People also waste water by watering lawns in places where they shouldn't have been grown in the first place. A lot of us are thinking right now that everyone is wasting water, watering lawns, filling up swimming pools, etc.

But we're wrong; most people are doing everything they can to save water. They made technologies where you can turn pee into clean water and ocean water into drinking water. Farmers are finding their own ways to save water. The state government and the people of California are setting goals and a lot of people are doing everything they can to meet that goal. However, for us to do better, all Californians should engage in this effort by not wasting a lot of water.

A lot of people keep saying that the farmers are wasting tons of water while California is in a drought. Well, here are some ways that farmers are trying to save water: drip irrigation, capturing and storing water, and dry farming. For dry farming, farmers "rely on soil moisture to produce their crops" according to the Center for Urban Education about Sustainable Agriculture. This helps the drought because they are using the water that is already in the soil. Secondly, for capturing and storing water, farmers "have built their own ponds" (CUESA) using their own systems and their own water that they have captured. Lastly, "Drip irrigation systems deliver water directly to a plant's roots, reducing the evaporation that happens with spray watering systems" (CUESA). This system occurs underground, and it doesn't spray water all over. In biological terms, these are examples of mutualism: farmers and the rest of us are all benefiting. These are some of the ways that farmers are doing everything they can to help us and themselves in this crisis.

The most useless water might be the most helpful water; one day we might have to drink our own pee. The San Francisco Public Utilities Commission has an "idea of purifying and drinking waste water—which includes water from your toilet," according to *popularmechanics.com*. You might think that's disgusting, but this is what's going to happen when we are really lacking water.

People also don't think they could drink salt water. Well, Santa Barbara is making "useless" salt water into drinking water. In this process, "two gallons of ocean water go in; one gallon of drinking water comes out. The leftover gallon contains super-salty brine," according to *npr.org*. This is a good idea because you can easily get salt water here in California. This is another example of mutualism; the people who make the useless water into drinking water benefit by making money, and the ordinary people who drink it benefit by not having to use up all their water resources.

Even though people say Californians are wasting a lot of water, most Californians are trying to save water and we actually met our goal and more. "Californians reduced water use by nearly twenty-seven percent during August, exceeding Governor Edmund G. Brown Jr.'s twenty-five percent conservation mandate for a third straight month," according to *ca.gov*. This is surprising because a lot of people think we waste tons of water in the summer because of the hot weather.

Farmers are saving water, people are developing new technologies, and people and the government are working together. But despite all these good things, it's still not enough. We may have reached our carrying capacity in California. Then, the tragedy of the commons is happening; some parts of Southern California are increasing water use. "About a third of water districts, 140 in all, fell short [of meeting the water-use goals], mostly in Southern California," according to KQED. This helps no one. If some parts of California keep using too much water, in the end we're all going to suffer and destroy our ecosystem. This is an example of competition for resources between Southern and Northern California, and in the end Northern and Southern Californians will both not benefit from it.

We can get out of this calamity if we all work together, just like how some farmers and people are doing. We just have to talk it through. Being greedy is not the answer in this kind of situation. In the end, if we don't work together, all Californians will suffer from this and we will affect other animals and ecosystems. We are doing great, but we need to do more because we're still in this in this calamity. Sometimes we just need to open our eyes and be mindful of what's happening around us.

Asante Lewis *is a sophomore at John O'Connell High School. He commutes from Vallejo, rising very early every morning to get to school. After school he plays baseball,* Call of Duty, *and hangs out with his friends. He is interested in one day going to law school.*

THE PIECES OF A PUZZLE

Sports teams are very similar to ecosystems. There are both biotic (living) and abiotic (non-living) factors on a baseball team. Some of the abiotic factors in baseball are: bats, gloves, cleats, and the field. These are important factors because they are key parts of baseball. Most sports rely on abiotic factors for the game to function, but there are also other biotic factors in sports such as the crowd, family, and friends. These other factors can also play a big part in how a game will end. There are also many different kinds of personalities, backgrounds, attitudes, and levels of morale. When these differences are put together on a sports team, it creates diversity. Diversity is important to a sports team because it can lead to success, create leaders, and teach social skills to players.

When a diverse team is put together it has a great chance of leading to success. My baseball coach says, "Putting together a baseball team is like putting different pieces of a puzzle together." We use communication to help us reach our common goal to win, which is also similar to mutualism in an ecosystem.

Diversity comes in many different forms, such as skill and personality. Different levels of skill cause a team to be strategically put together. Teams with a diverse skill set push each player to excel at one's own role. When I see our starting shortstop play very well consistently, it encourages me to be better and play like he does. Similarly, I am more of a defensive player and hitting is not what I'm known for; I'm known for speed and a strong arm. On the other hand, my teammate, Nick Brown, our catcher, is known for hitting and is not considered one of the fastest players on the team. So when a team has different skill types like this, it makes the team more flexible and diverse and can drive the players to be more successful.

There are also personality differences on a team that make it diverse. For example, my teammate, Gabriel Rojas, is a lot more calm in morale and attitude, whereas I am much more vocal and emotional when I play baseball. He keeps to himself, and I usually let out how I'm feeling. He is also a leader on the team. The two different personalities or attitudes benefit the team because it allows character and bonding to happen, which overall increases the team communication. It may sound weird to have that much of a difference between players, but that's what makes it a team.

In baseball, players tend to be altruistic, like when a teammate sacrifices themself so another teammate can advance to the next base. Also, a major part of the success of any team is picking up things from teammates or learning from each other. When my teammate, Gabe, uses a specific form or I see him make a nice play, I study him and try to do it myself. Also, my teammate, Javier, says that when he sees me make a mistake, he learns from me and tries not to do the same thing. When Javier learns from my mistake, this is a form of commensalism because when I make a mistake, Javier gains a positive lesson and that doesn't affect me negatively.

When a team works together to complete the common goal of winning, it is a form of mutualism. This is how a team gets better: by learning from one another and improving on its weaknesses.

Some may argue that a team with similar traits leads to better success than a diverse team. This is sometimes true when players grow up together and have long-time bonds. However, there are still plenty of teams that follow a different path. Similarity on a sports team can cause friction and selfishness, while diversity causes teammates to learn and understand each player's standpoints. For example, Shaun Livingston plays for the Warriors and is a veteran player in the NBA. He most likely shares knowledge with his younger teammates. In return, the presence of his younger counterparts gives him drive and energy to play with their level of enthusiasm. This shows that diversity on a team like the Warriors causes great success in the trade-off of energy and wisdom.

In conclusion, diversity is a very important factor for ecosystems as well as sports teams. Diversity on sports teams and in ecosystems make the two complete. Each needs different parts in order to produce the most successful outcome. Diversity in ecosystems can be applied to many different environments. There are other connections in life that you can make with diverse individuals. When diverse people work together, it's a mutual learning experience.

Maria Verduzco *was born in San Francisco and enjoys listening to hip-hop. She likes to dance and roam the city with her friends, going to parks, the pier, and the movies. Her parents are good cooks and she loves to eat their spaghetti, chicken, and rice. After she finishes high school, she would like to go to college to become a pediatrician because she loves to take care of children.*

STORIES WE TELL

"We might be entering into a kind of community crisis," says Fred Blackwell, Chief Executive Officer of the San Francisco Foundation, a community advocacy organization. He said this because San Francisco is becoming less diverse. The white population in San Francisco is on a steady increase. Just like biodiversity helps an ecosystem because diversity helps to keep all the living things alive, people from diverse backgrounds can learn from each other. Everyone should get to know each other so everybody can tell their diverse stories. They could learn how other people cook, and talk to other people. Communities that don't have diversity have a problem, because if you don't get along with people who are different from you, you're basically the one who is suffering.

My parents talk while they are cooking dinner. While they are cooking, I am listening to music. My brother watches TV. Usually, when dinner is ready, they call my brother and me to the table. We sit down and we eat. That's it. We eat in silence. Everyone does their own thing while we eat dinner. I listen to music and my brother watches TV and my parents talk.

One night I wanted to break the quietness by having communication with my parents so it could be noisy. I was thinking that it might be more fun if there were more people because we could talk and ask each other, "How was your day?"

I like going to family gatherings because it's noisy. It's fun and everybody can talk about their lives and what they are doing. When I and my cousins talk, we talk about life and where we've gone and stuff related to that. While we talk, the adults talk or make food.

In this way, my family is like an ecosystem. I believe that if there were more biodiversity in the house or in the community, our ecosystem would be better. It would be better because people could work together and communicate. The communication would be about their lives and their backgrounds and how they got here. The more interactions people have, the stronger the ecosystem or the community can be.

I decided to talk to my parents so it wouldn't be quiet in the house anymore. That's how I tried to make my house diverse. In an ecosystem, increasing biodiversity means that the ecosystem would be stronger. Similarly, increasing diversity in a community can mean more interactions or connections made between people. When people interact with each other it makes them feel better and important.

I was curious about how my family got here. One day before dinner a long time ago, I asked my parents how they got here. They said that when they came to the United States from Mexico, they knew it could be dangerous and anything could happen, like running out of food or water or not having enough sleep.

I was shocked because I didn't know that a lot of people would go through that just to have a better life.

It was hard, they said. People told them to get in cars so they could cross the border. They didn't know who they could trust. The trip was full of hidden dangers, so all their senses had to be alert. Just like prey in an ecosystem, they had to watch for predators that wanted to get them.

They weren't the only people in the back of the truck. There were a whole lot of people so they had to be still. They knew they were safe on the other side because the people in the front seat of the car told them, "We passed the border." When they finished telling me, I said, "Oh." But I felt sorry for them and all the people who have gone through that. I couldn't express how I felt. At the time I didn't say, "Oh are you for real?" or anything like that. I wouldn't want to explain how I feel because it's just sad.

It's nice in Mexico when we go see our family who lives in Michoacan. It's hard to get jobs there, though. Here in San Francisco there are more opportunities, such as jobs and education. There you can go out and no one cares when you come home. You don't have to worry. Everybody knows each other. It's not diverse, but everybody knows each other.

Here in San Francisco, within your community, it's not always diverse either. You don't know everybody in the city, but you know the people that live around you. At this time of my life, I prefer to live in a community that is not diverse, but later on it would be nice to live in a place that is diverse because you could get to know other people's background stories and learn more about them.

I was glad I broke the silence in our family because we talked and I had an opportunity to talk to them about their past. I haven't gone through what they did so I don't know how it feels.

People should think about how to make their communities diverse. If they do, they would like it because it's nice to see people around you and to learn from each other. Everybody with different races should get along because what's the point of not talking if we're all here? I recommend you to try it with your family first and then with the people around you. This would make your ecosystem more successful.

MY CULTURE IS NOT
YOUR COSTUME

Change, challenge, and holding on to roots

Vanessa America Ramos *is a strong, confident fourteen-year-old freshman. She is a culmination of many cultures and identities. Her mother is Salvadoran, Cherokee, Slovenian, and Italian. Her dad is Mexican and Colombian. She sports turquoise hair that matches the ring her mother gave her. She works at Roadmaps to Peace, which occupies her afternoons. She hopes to work in juvenile detention as a probation officer so she can help young people.*

THE "NEW MISSION"

Walking in San Francisco's Mission District is very different now than it was before. When I was younger, I only saw people of my own kind (Latinos). There was a lot of cultural food, Mexican and Latino places. Now these businesses that have been around for years are getting shut down and being replaced with high-tech places and condos. The type of people I see are mainly Caucasian, higher-class people that live in luxury apartment buildings, where before there were regular houses. Now people who have been living in their homes for years are getting kicked out. The Mission is changing because of gentrification; the social community is changing because Latinos are getting evicted and people with money are coming in.

Personally, not much has changed on the block that I grew up and live on, but walking around my neighborhood I see all of the changes that I listed above. Walking around my neighborhood, I used to feel comfortable and at home. Nothing really felt out of place before. Now walking around, even to go to the corner store, I feel looked down upon.

In the middle of elementary school, things started to change and not necessarily for the better. I don't remember speaking about this topic or learning about how gentrification is affecting us big time, but I grew up and started to see everything for myself. I noticed that things weren't right and saw a change.

The Mission is a changing ecosystem day by day. Money, limited housing, and the culture are just a few examples of abiotic factors here in the Mission. There are also many biotic factors that interact with each other such as native people, the incoming people, and the landlords. The non-white population in the Mission dropped from 71.8 percent in 1990 to 57.3 percent in 2013, according to the United States Census.

I talked a little bit about population, now I am going to talk about the money, which is an abiotic factor. The average for a one-bedroom in the Mission is $2,600 per month, according to Zumper. Also on Airbnb, the average cost for a night is $203 to stay in the Mission. According to Zillow, a two-bedroom, one-bath apartment is $4,500 per month. For example, I even know a college student who is having financial troubles. He lives with three generations of his family—four people in all—in a one-bedroom, rent-controlled apartment. This month, his landlord increased the rent to $1,700 a month from $1,500 after not raising the rent for several years "We're barely surviving," he said.

If the Mission is the ecosystem, the carrying capacity is the maximum population size in the Mission, and the predator/prey relationship is when the landlords act in their own best interest, but at the cost of everyone else's success.

The culture is changing as the new incoming residents arrive. Martina Ayala, a teacher, artist, and consultant for San Francisco nonprofits working with low-income families, says, "They're trying to get us out without having to pay the eviction costs. And so they're doing that by harassing us and calling us every day, sending us a three-day notice to pay rent or quit without following through with service. Even though we are paying $1,750, that is still not enough for the landlord, because the average rent is now $3,000."

There's a lot of biodiversity in the people living here in the Mission. There are different ethnicities of people walking around, and there's not only Latino cultural food, but also diverse food. Equilibrium is disrupted in this social community when natives are not having any say in what's happening to them and their homes. Parasitism in the Mission, for example, is when someone who has been living in his or her home for a long time and is highly affected by the incoming people gets evicted and is not left with much.

The Mission has competitive ecological relationships. If the people want a house, it's going to be a competition of who has more money because everybody wants to live there. The equilibrium is maintained in this social community when the natives fight for what they want and take action for what they believe in.

There are many differences in the Mission now. For example, the food is different, like the pizza places on 24th, and there are new buildings. The landlords are not being caring. This is causing a lot of families to get evicted and move out, and this causes wealthy people to move in because they have the money for it. It's not the same culture. For example, the Day of the Dead is a part my culture. When my mom first started to take my siblings and me, it was to celebrate the death of my loved ones. Now I feel people are coming from different parts of town to just party and dress up, but they don't know the meaning or where the roots come from. They dressed up because they thought it was a costume to wear, but my culture is not your costume.

Eneiri Jocelyn Serrano *is interested in big questions —like where do we come from? And how do human bodies work?—which is why she enjoys biology, world history, and reading nonfiction. She aspires to be an investigator with the FBI. An avid soccer player, she also likes indoor rock climbing and horror movies.*

STOP DISSING MILEY CYRUS; CULTURAL APPROPRIATION IS NOT AS BAD AS YOU THINK

I go to a school where there are students from many different cultures, but we like to mix it up. My Honduran friend likes rap, and likes to wear fashion borrowed from African American rap artists. My Latina friend likes to wear big hoop earrings which originated from African American culture. And I like to dress in many ways, so I mix a lot of different cultures at the same time. Recently, for my fifteenth birthday, I got a tattoo. It was borrowed from many other cultures. I don't even know where it started.

Some people say the borrowing of cultures is cultural appropriation, which has been controversial. In the news recently, Miley Cyrus was criticized for twerking, Karlie Kloss wore a Native American headdress as a Victoria's Secret model, and Rachel Dolezal, a white woman, passed as an African American in order to lead the NAACP. These are all examples of cultural appropriation done badly, because they trivialize other peoples' culture. But I think cultural appropriation can be positive because it can strengthen the community. In this way cultural appropriation is similar to biodiversity in nature. Biodiversity refers to the variety of life on the planet, including plant and animal species. In nature, biodiversity is important because it creates a healthier ecosystem.

Cultural appropriation is the same thing. By definition cultural appropriation is the use of elements of one or more cultures by members of a different culture. At first, cultural appropriation seems bad, because it can be demeaning to minority cultures. Going back to the previous example of Rachel Dolezal, would you want someone to pass as your color? That's what Dolezal did, and a lot of people were offended by it.

But cultural appropriation can also have a positive impact. Appropriating fashion can be positive, depending on how you wear it, use it, or represent it. Fashion is not only fun but it can be a way of sharing cultures. It's not just fashion that is being appropriated; here in San Francisco's Mission District you can see examples of cultural appropriation in the variety of foods, styles, and languages that people borrow from each other and share.

I think that this kind of cultural appropriation strengthens my community in the same way that biodiversity strengthens the ecosystem. I feel that cultural appropriation makes us more social, friendlier, more understanding of each other, and more willing to talk to each other. I know that this has happened to me. I've learned a lot from interacting with people from other cultures and it has made me a better person. The more we know about each other, the stronger we will be. Instead of having so much discrimination and racism, we can support each other through rough times.

In nature, biodiversity is an advantage because it strengthens the ecosystem. In our community, cultural appropriation acts as biodiversity, and we are stronger for it.

Cesar Diaz *was born and raised in San Francisco in the Mission District. He enjoys going out in the city on his bike. He has plans to study business and start his own company someday. He collects interesting shoes and is interested in textiles and construction.*

ON A MISSION

According to the Merriam-Webster dictionary, "Gentrification is the process of renewal and rebuilding accompanying the influx of middle-class or affluent people into deteriorating areas that often displaces poorer residents." Gentrification is a difficult topic for a boy who's living in San Francisco's Mission District, where this is happening. It is very sad knowing that stuff is being changed. Although change isn't all bad, it's good only if it's good for the people.

The Mission District was all once saltwater marsh. Then humans started putting in dirt to cover it up to make more land. Apparently, in the 1700s a Catholic mission was built on present day Dolores Street, it still stands as the oldest building in San Francisco. And most of the dry land around the mission was for rancheros working for wealthy Mexican families. Then, with the Gold Rush, immigrants rushed in from Irish, German, and Italian backgrounds. Then the great earthquake and fire of 1906 terrorized the neighborhood. Later, in the 1940-1960 era, when the Bay Bridge was being built, Latino communities were forced to move from the area around the Bay Bridge and into the Mission, which was the beginning of the culture that thrives here today.

Now that you know the history of the old days and why there is a large Latino community in the Mission District, we can talk about how all that is fading away due to gentrification. Here in the Mission District there have been multiple cases where people have been kicked out of their homes. According to *The New York Times* article "Gentrification Spreads an Upheaval in San Francisco's Mission District" by Carol Pogash, "When a family in a rent-controlled apartment leaves or is forced out, the rent is jacked up to market rate." This shows that families are forced to leave their home because they do not have enough money to pay rent. Additionally, she writes about Paula Tejeda, a Mission resident who owns an empanada shop: "Her shop could not endure years of construction and the loss of customers, many of whom she knows well enough to greet with a kiss on the cheek." This shows that Ms. Tejeda had a friendship with her clients, and then all of sudden, a 331-unit apartment building was built in front of her store, which influenced the people around her, added new faces, and interrupted her business and her friendships. This represents disruption in the Mission ecosystem because Ms. Tejeda's business is getting hurt because this new building was being built. This is very important because the new things in the Mission are affecting her business, which will not be good for her.

Gentrification can be a heartbreaking subject topic to talk about. Some could say that you really can't say when gentrification is happening because there are not enough facts, but my personal experience is a fact to me. I have seen gentrification happening all around me here in the Mission District. The One Dollar Store building at Mission and 22nd Streets was rebuilt into some apartments with a stupid window design. If you walked by there, you would think, *this building looks out of place. This does not belong here.* It looks like it doesn't belong on a street where there is a Latino vibe. Also, a building next to it was burned down "by accident." My mom would always go there to get groceries. All of a sudden an old building gets burned so they want to build a "better" place, which just means benefits for faceless companies.

I was born and raised in the Mission District. During the summer as a kid I would go to the store and see these old buildings that made me smile. I'd see all the African American and Latino people, and now all of sudden I see more whites. I am not being racist, I don't mean it like that. I'm just saying that things change and it affects the community. Jennifer Perez, a friend of mine from the Mission, says, "I've seen changes here in the Mission District. [I know it's] something called gentrification."

Another one of my closest friends, Marco Marmol, says, "I think it's messed up how they're making the prices so high that people get kicked out of the place they've called home. I've seen a lot of changes in the Mission District, like new apartments being built."

All of this is tied in with gentrification. Most of these people who get kicked out have no other option but to leave because landlords raise their rents without caring for anybody or anything but the money they're making. After somebody leaves, who moves into the building? A family or a person who has a lot of money and a great job.

Gentrification can be related to ecological relationships like predation. My tutor Marcus said he lived in an apartment in a small house. This building had seven tenants. People were moving in and moving out all the time because the rent was so expensive for the apartment they were living in. There was a lot of pressure from his landlord, who did not live in the city and kept raising the rent. The landlord had made an enemy with a neighbor when the neighbor was building a new house, so paying so much for rent felt empty because it wasn't a safe and positive place to be. So then he moved out. This is parasitism because there was one person being benefited, which was the landlord who was making more money, and Marcus was being negatively affected by losing money. The people who are coming in with money and investing it in a building and making it brand new are like the predators, and the people who are getting pushed out of their home are the prey.

In conclusion, many things have happened due to gentrification, and we cannot stop it since it will happen either way. We can protest and we can riot, but nothing will stop this cycle. All we can do is make people aware of gentrification and how it is changing lots of things; it's kicking people out of their houses, it's causing stress on families. Here in the Mission District we just see things changing. Something I want to do is be on a mission to keep the Mission the way the Mission is supposed to be.

Maika Suzuki *was born in Sendai, Japan. She moved to San Francisco four years ago with her mom and dad. She likes to look at the comics and manga in the bookstore in Japantown. Maika is the author and illustrator of a comic novel about a martian. In the future, she wants to write more comic novels about space travel.*

HOW MY FAMILY SURVIVED THE GREAT EAST JAPAN EARTHQUAKE AND TSUNAMI

I felt the earthquake that caused the tsunami off the eastern coast of Japan on March 11, 2011, when I was ten years old. It was a Friday afternoon. I was at school in Sendai. It was a 9.0 earthquake that lasted five minutes. It was two earthquakes at the same time, and together they are called The Great East Japan Earthquake. They triggered the tsunami that followed.

Everyone in my classroom was screaming and crying when the earthquake happened. The teacher told us to go under the table. We were very scared because the earthquake was the biggest one we had ever felt. While the city was shaking, my dad came to my school. He ran into the classroom and he picked me up. The minute my dad picked me up, I felt safe.

My dad brought me home and we waited for my mom, who was shopping with her friends. While I was waiting for my mom with my dad, it was very cold outside. We looked at the sky and it was snowing! It never snows in Japan in the spring.

Then together we drove to grandmother's house in Taihaku Ward in Sendai. When we got there, I listened to the radio and they said that 2,000 dead bodies were in the ocean because of the earthquake. When my mom came, she said that three days before the earthquake, she saw that it looked like there were two suns in the sky. Two suns in the sky is a sign that an earthquake is coming.

My mom's older sister's family was already at my grandmother's house when we got there. It was dark in my grandmother's house because she had no power in her neighborhood. We didn't know about the tsunami yet. We called my father's family and they said my aunt and uncle were missing in the tsunami. After the tsunami, all the ships were broken and many dead people were in the ocean.

My dad said he wanted to help people in Ishinomaki, where his family lived, about two hours away by car. People lost families and houses in the tsunami, so he wanted to give food and blankets to the survivors. He also wanted to find my aunt and uncle, who were missing. In an ecosystem, when both members benefit from something, it is called mutualism. By going to Ishinomaki, my father was helping the survivors of the tsunami to get food and he was also searching for members of his own family.

I waited for my dad with my mother's family at my grandmother's house. I have a cousin, Miyu, who is the same age as I am. We were scared that more earthquakes might come to our city, Sendai. To try to keep our minds off the tsunami, we played video games on our DS devices for about ten hours, until the batteries went dead.

We stayed at my grandmother's house for one or two weeks and then the power and the water came back on. My father has many friends and they shared their food and water with us. We couldn't use the bathroom in my grandmother's house because the water had stopped. So we melted snow and used it to flush the toilet when we went to the bathroom. When my dad came back from Ishinomaki, he said he couldn't find my aunt and uncle. He told us that his cousin and her babies were dead. "There were so many dead bodies and rubble on the street," he said. "They found my cousin and her two babies, but they did not find your aunt and uncle."

The tsunami disrupted the fishing industry in Ishinomaki and it also disrupted the life of my family. Humans are helpless in a tsunami, all you can do is run or drown. A tsunami is a human and ecological disaster.

The population of Ishinomaki is 163,000. The death toll from the earthquake that caused the tsunami was 3,162 and 430 missing. Twenty thousand people lost their homes. The highest crest of the ocean during the tsunami was sixty-six feet, but it also reached phenomenally massive heights of 131 feet.

Outside in Ishinomaki, there were mountains of rubble. Cars were stacked inside houses. There was an oil tank on top of a house. Survivors were rescued in boats. In Sendai, where I was with my family, only some buildings collapsed. There was not as much damage as in other places.

When we went home from my grandmother's, the power was on but my room was a mess. Shelves had fallen. Windows had cracked. Books were scattered all over the floor. My mom, dad, and I got to work. It took a day and a half to clean it up. Cleaning up was troublesome. It was a lot of work.

Two weeks later I went back to school. There was not much food during the disaster. The school's lunch facility was damaged during the earthquake, so when we returned, school lunch was only bread and milk.

There was another earthquake in April when I was at my grandmother's house. I was sleeping. It was midnight. I didn't know what had happened. It was strong, but I didn't feel it at all. My grandma woke me up after it was over. In my family, I was the only one who didn't wake up. My cousin laughed at me. She said I was stupid. I did feel a little foolish, but I explained that I felt sick before I went to sleep. In a way, I wish I had been awake for the earthquake, because then my family would not have called me stupid.

About three months later, in the summer, a dog came to grandmother's house. Kiki was the name of the dog. She was a female. She was a childhood friend of Pori, the dog my grandmother lost the year before. When Kiki came to my grandmother's house, her owner had been away from her house for about one month because it was in a danger zone. Kiki couldn't eat much food during that time, and she was so skinny. My grandmother was happy that another dog came to her house. My grandmother also has two cats, Meru and Kuu. Five years later Kiki, Pori, and her son Maru, and my grandmother's two cats died. Papi is the only dog left in her house now. She also has her two cats, Ran, and her daughter, Chii.

I am glad that I live in San Francisco now. I love animals. I would love to have a cat, like my grandmother in Japan, but I have two hamsters at my house so I can't get a cat. The ecosystem in San Francisco feels friendlier than the ecosystem I lived in in Japan. In San Francisco, I am not as scared of earthquakes as I was in Japan.

Rigoberto Sanchez *(whose full name is Rigoberto Guzman Sanchez, Jr., but call him Rigo), is a proud San Francisco native with two sisters and a brother. He works part time at a day-care center and enjoys playing with his little nephew. He likes his history classes the most and wants to learn more about politics. Rigo also enjoys working with his uncle on cars, and thinks he'd like to be a mechanic someday.*

THE OLD PLACE I LOVED

My research shows that there are big changes happening in San Francisco's Mission District, and that those changes are affecting natives and long-time residents. The Mission's social community is like an ecosystem in that the new people moving into the neighborhood can be seen as an invasive species, displacing residents who have lived here for a long time. Like a natural ecosystem, these changes threaten the biodiversity of the social community. You can see this because my research shows that there has been a big shift in race, income, and housing costs in the Mission.

Growing up in the Mission, I've noticed multiple families moving out of the neighborhood. Even my sister is planning to move forty miles away with her baby because she can no longer afford to live near where she works. It has me thinking that someday I may have to move away from the city where I grew up.

Not only are families affected by rising rents, but longtime businesses like shops and restaurants are being forced out by higher rents. They are being replaced by businesses that cater to wealthier people. This leads to resentment and a fractured community.

It's clear that there's been a big shift in the demographics of the Mission District. There are more Hispanics leaving and more Caucasians moving in. Because housing is at its carrying capacity, there is competition for that housing among the people who can afford higher rent, which displaces old residents. That means housing is becoming less affordable for the people who already live in the Mission.

The newcomers have a much higher income than the longtime residents. Census data from the American Community Survey shows that between 2000 and 2012, the white population in the Mission increased by twenty-five percentage points, while the Hispanic population decreased by ten percentage points. In 2010, the survey showed that whites surpassed Hispanics to become the majority race in the Mission District. The American Community Survey showed that white households in the area now earn a median income of nearly $37,000 more than Hispanic households.

A survey done by *The San Francisco Chronicle* showed that the median rent for a one-bedroom apartment increased drastically between 2011-2015, from $1900 to $3610. That's a ninety percent increase, the second highest of all neighborhoods in San Francisco. With these very high rents and competition for housing, it's not a surprise that many longtime residents can no longer to afford to live here.

The statistics in the *Chronicle's* survey are supported by my own informal poll of ten longtime teenage residents of the Mission District. I asked them, "Have you noticed changes in the Mission District over the last several years?" Every single one said they have noticed changes in the neighborhood. They all noticed new people moving in, some longtime residents having to leave, and new shops and restaurants displacing the old ones.

To look at this issue more closely, I interviewed sixteen-year-old Ricky Gallardo, who has lived in the Mission District his whole life. "The change has affected the Mission, and I can't go to some of the old places I loved… families are getting pushed away from the Mission, where they grew up."

These major changes in the social community, which is the Mission District ecosystem, lead to resentment and a more fractured community. This can be seen as a predation, because wealthier people who see the Mission as a nice place to live may be in some ways creating hard feelings which eventually hurt the natives in the community.

There are changes that can help the Mission and businesses. We could build more affordable housing in the Mission, lower the business taxes for existing shops and restaurants, and enforce more rent control for landlords who own buildings. These are some ways we can slow gentrification down in the Mission and in the city.

Oscar Moreno *enjoys playing soccer, hanging out with his friends, and eating (he'll eat anything at all.) At the age of sixteen, his plans include going to college and getting a good job – a job that is not in an office, and lets him work with people. He's okay if the job includes math, but only if the math is counting money.*

DIVERSITY MATTERS: RACISM FROM THE HOLOCAUST TO TODAY

In my own life, diversity matters. It matters because people could look at me in a different way than they look at people of other ethnicities, such as people of European descent. It would make me feel disappointed if I were looked at differently, because we are all human beings who live on the same earth, in the same galaxy, in the same universe.

Diversity matters not only in society but also in ecosystems, because everything is connected. I don't understand why some people don't value diversity in life.

When I read the book *The Boy in Striped Pajamas*, I was inspired to write about the Holocaust because I was shocked at what I learned. It made me wonder why innocent people were being killed. It was painful to read about a young boy who was hiding out from the Nazis during World War II, and then was later captured and taken to a camp, where Jews were killed and he was also killed. The showers had poisonous gas, which is how the boy was killed. There were other kids who actually escaped the concentration camps, but were caught and returned to the camps where they were tortured. I couldn't understand why there was so much hatred for no reason.

Ironically, some people think Hitler himself may have had Jewish ancestry. Whether he did or not, why would he kill people simply because of their ethnicity? Not only were Jews affected by the Holocaust, many others were also affected. Anyone who didn't fit the Nazis' specific expectation of how you should look or who you should be was affected. Approximately eleven million people were killed during the Holocaust, and six million of them were Jewish, according to the United States Holocaust Memorial Museum.

Hitler was a predator. The Nazis were trying to take over the whole world. Some of the Nazis actually enjoyed doing what they did; others were forced to do it. Not only were adults killed, but little kids and babies were also killed. The Nazis considered kids "useless" because they couldn't work. It is estimated that 1.5 million Jewish children were killed, whether in the gas chambers or shot in front of ditches dug for mass graves. Older children survived by being forced into hard labor. Babies were killed instantly. Other kids were tested or used for experiments, according to the United States Holocaust Memorial Museum

This history makes me feel sad. It's just messed up. Why did all of this happen? If I was there during that period, I'd feel scared. In the present day it makes me mad because people are still getting killed because of how they look or who they are, when they were just people who wanted to live their lives. Now this is happening with police brutality toward people of color. I know many people who have been arrested simply because of their skin color. As a boy, I even knew people who were killed on the streets, or randomly arrested for no reason.

Today people are shot for resisting arrest. What the boy in the book I read saw were people crying, screaming, and scared. Some kids even saw their own parents die. It didn't surprise him because he'd seen it before. Right now teenagers are getting shot. I know this is happening, and it makes me feel mad.

In ecology terms, biotic factors are living things and abiotic factors are nonliving things. In the time of the Holocaust the biotic factors could be seen as the people and animals. The abiotic factors were the camps, the poison, and dead bodies. It would be the same today, biotic as the living people and the environments in which we live, and the abiotic factors as weapons and technology.

Since the ecosystem and integrated human community was destroyed by the Nazi takeover, the safest place to be during the World War II period would have been in basements, attics, or secret rooms behind bookshelves. Many people left their countries and survived this way. Today, I think the safest place to get away from the brutality in society would be your house or somewhere outside the city, like on a farm.

Across history, many people want to maintain the interconnectedness of all people, no matter their skin color, and of life itself, even when other people are trying to disrupt and destroy life. For example, according to *history.com,* there were the "freedom riders," a group of thirteen African Americans and some white Americans in the early 1960s, who were peacefully protesting segregation in interstate bus terminals and were bombed. And Martin Luther King, Jr., who was protesting, advocating for civil rights, and trying to convince people that African Americans deserved equal rights and the right to a peaceful life, just like anyone else. He was assassinated in a brutal way for standing up. And then there is the Ku Klux Klan, which is known for chasing after African Americans and burning down their houses, so people have to escape to save their lives. There are both innocent people and psychopaths throughout history. It's the same pattern. In ecosystems everything is connected and interrelated. Human violence affects us all, whether directly or indirectly.

Imagine a forest. There are a bunch of redwood trees, and then there is an oak tree. Someone wants to cut the oak tree down because it is different from everything else. And then it isn't there anymore. Unless, of course, you plant another oak tree.

Imagine a person, a Latino who grew up in Mexico, who experienced gang and cartel violence, and who moved to the United States for safety and to find a better life. Now imagine everyone around him comes from a different country. If the Latino was kicked out of that space, the other people would miss out on hearing his views, experiences, and perspectives.

Imagine a teacher. One teacher could know more about history, and another teacher could know more about something else. If they fire the teacher knowledgeable about history, the students would miss out.

I am trying to express why diversity is important. It is important because the many ways that people see life all matter, and to lose their voices, experiences, and views is a great loss. It also affects history, depending on who tells the stories. The book *The Boy in Striped Pajamas* is a true story told through the perspective of people who endured the Holocaust firsthand. Without the Jews, we wouldn't know a lot of their history, the experiences they had during the Holocaust, and the brutalities they suffered.

If everybody were the same it would be very awkward. If everyone looked exactly like you and had a very similar background, it would be really boring. It would be the same as if a forest had all the same trees, as compared to a forest with hundreds of varieties of trees and plants, animals, and birds. Diversity matters not only in my life, but in the world, the universe, and beyond.

Briana Martinez *is a fifteen-year-old freshman at John O'Connell High School. She was born in San Francisco, but is fortunate to frequently visit her family in Mexico. When in Mexico, Briana especially enjoys going to the fiestas de Diciembre and fiestas de Agosto, where the entire town has a party with firecrackers and a crown falls from the sky to be caught by a lucky person.*

MISSION TRADITION

To some people the Mission is just another district in San Francisco— just another name, just another place to buy souvenirs. But in reality, it's a home to me, my family and many others. The Mission has always been a place heavily based on community, family, and culture. Because of gentrification, the equilibrium in this community is being disturbed. Many people have lost their businesses, homes, and daily routine.

When there are eviction notices on people's front doors and landlords decide to build apartment complexes, the people moving into the Mission and entering those new homes are surrounding themselves with something new and strange to them. This new thing is special, unique, and what sets the Mission natives apart from other people. It's our culture and the roots that we have deep in the ground, which are getting cut as the people keep getting evicted so more wealthy people can move in. People with money and power have always done things without considering or caring about the people they affect long term.

My family has lived in the Mission District since 1960. We've seen so many people come and go, it's starting to become normal. The Mission is now changing at a very fast rate. They're making homes that regularly accommodate two families into very luxurious homes that are only available to one family. This makes it extremely difficult to house as many people as possible.

The huge apartment complexes are primarily for the wealthy because they are the only ones that can afford them. The Mission has generally been for the middle class, but in today's world it's difficult to live here on that wage. According to "How Much of Your Money Should You Be Spending on Rent?" by Sienna Beard, "You shouldn't spend more than roughly thirty percent of your income on rent." In that case most of the people currently living in the Mission are not living in the right place. According to *SFGate*, the average person in San Francisco spends closer to fifty percent of their income on rent. All people are affected by this. This includes our teachers, firefighters, construction workers, and pretty much anybody that makes up the middle class, which makes it harder for families to find a good home.

The 2014 Day of The Dead parade was where I really started to notice the change made by gentrification. As I watched the parade, I saw many people that looked like tourists, interested and amused, taking it in without benefiting us, like the invasive species in our ecosystem. They weren't fully understanding the meaning of the day. In some ways they made a mockery of a sacred tradition. They saw it as a costume party, which it is not. The Day of The Dead is a time when we celebrate the lives that our past family and friends lived, a time when they can join us, so they know they will never be forgotten.

To fully comprehend what is occurring, people need to become more aware of the people they are hurting when they move into a neighborhood. I don't blame them for not understanding our culture. I feel all they want is a part of our culture, but our culture isn't something you can choose or something you can just pick up. It's something you're born into. It's a way of living. All they're really doing is participating in our way of living. They see something they like and they try to take it with them. They see something they don't like and they just replace it with their own thing, or just remove it.

Maria Novelo *comes from Mexico and has two main passions: dogs and make-up. She started getting into make-up when she would apply it for her mothers' friends—she wasn't allowed to wear it when she was little, but now that she's a teenager she can practice on herself, too. She hopes to be a professional make-up artist someday.*

PUNISH THE PEOPLE WHO POISONED FLINT

Imagine going to take a fresh shower, and when you get to the bath and turn on the water, the water is dirty, brownish, and has a nasty smell. How would you feel if this was your situation? What would you do?

One time I was going to take a shower, and before I opened the water faucet something nasty came up from the bathtub drain. The water looked gray and black, like dirty water from the washing machine after dirty clothes get washed and all of the dirt and hair goes into the water.

The water smelled a little weird, like metal. I had to put my clothes on again to go tell my mom what happened, and I couldn't take a shower that morning. I felt like, *What happened here?*! I never saw dirty water coming out from the pipes before. The dirty water lasted one week. I felt upset because I couldn't take a shower because the bath was full of the dirty water. I needed to go to our upstairs neighbor to take a shower. I had to ask her if she would let me use her bathtub to take a shower, and I didn't know how to ask. That was uncomfortable for me.

I can imagine how the people of Flint, Michigan, feel about the toxic water in their homes.

I felt upset and angry when the water was dirty in my bath because not taking a shower for days can make you feel gross, and your body can get infections. The people in Flint probably feel the same thing. They probably feel even worse, because they can't even drink the water because it is toxic and makes people get sick. "This could have been prevented if, in accordance with the federal Lead and Copper Rule passed in July 1998, the Michigan Department of Environmental Quality (MDEQ) insisted on implementing a corrosion control system when they switched their water source," according to *The Guardian*.

The people who are responsible for the toxic water in Flint, Michigan should be punished. According to a lawsuit filed by members of the community of Flint, the people responsible for the water problem are Governor Snyder, the state and city governments, and thirteen public officials. The MDEQ public officials who were supposed to check the water of the Flint River did not do their job, which was to treat the water with an anti-corrosive chemical so that the water would not corrode the water supply pipes. Because the water was not treated, the water did corrode the pipes, and the water became toxic with lead from the pipes, says CNN. Many people in the community started to get sick.

According to *The Guardian*, "Lee-Anne Walters and her family in Flint, Michigan, drank water laced with hazardous levels of lead contamination for nearly eight months beginning in the spring of 2014. The water was brown. Her three-year-old son, Gavin, broke out in a rash every time he had any contact with the water in their home. He would have clear water lines on his body after getting out of the bath. He stopped growing. The whole family broke out in rashes five times, and doctors treated them for scabies. On April 2, 2015, Gavin was diagnosed with lead poisoning. Today he is one of at least 27,000 children in the city who have been exposed to lead contamination, according to local news sources." CNN reported, "Dr. Mona Hanna-Attisha was seeing more and more worried parents fretting over rashes and hair loss."

Lead poisoning is a serious sickness. "Lead affects children's brain development resulting in reduced intelligence quotient (IQ), behavioral changes such as shortening of attention span and increased antisocial behavior, and reduced educational attainment. Lead exposure also causes anemia, hypertension, renal impairment, immunotoxicity and toxicity to the reproductive organs. The neurological and behavioral effects of lead are believed to be irreversible," according to *The Washington Post*. People with lead poisoning can get serious problems, and it's not a joke.

The city of Flint is an ecosystem. The community is a natural system made up of residents, which are biotic factors because they are living people and consumers, state and city governments, which are also biotic because they are made up of people, and the Flint River, which is an abiotic factor because it is not alive. The residents are interacting with the government by protesting and complaining about the toxic river, the toxic water going to their homes, the sickness they are getting from the toxic water, and the lead poisoning their kids are getting. The government is interacting with the residents by saying there is nothing to worry about. The residents and the Flint River interact when the people drink the water from the river and then get sick from the water, and become upset because the government is not doing anything to fix the problem.

Flint, Michigan, is an example of tragedy of the commons because the city government wanted to save money, so it decided to switch its water supply from Lake Huron in Detroit to the dirty water of the Flint River. The government acted in its own best interest by changing the water supply, but saving money came at the cost of the health of the residents of Flint because they got sick. This is an example of parasitism because even if the people of the government get sick from the water, they don't have to worry like the residents do about where to get money to pay for good water or for their medical bills. The parasite is the government.

The situation is also like predation. The government is stronger than the residents because the residents have to do what the government says even if the residents don't like it. This ecological relationship made it possible for the government to be a predator and save money by preying on the residents. But the plan to change the water supply source backfired because not only are the residents being harmed by the decision to change the water source, now even the people in the city and state governments who made the decision are getting hurt because they are getting taken to court, made to pay money, probably fired from their jobs, and their reputations are getting ruined. Some of the public officials might even go to jail.

The state and city governments have behaved selfishly with their city. They could have done a lot of things. For example, they could have treated the water with the anti-corrosive chemical, changed the water back again to Lake Huron in Detroit or they could have donated water bottles to the residents. But instead, their responses were to not worry about it, everything will be alright.

The best way to solve the toxic water problem in Flint, Michigan, is to switch the water supply source back to the clean water from Lake Huron in Detroit. Some things we can do to help the people of Flint are raise funds for clean water, donate bottled water so people can drink, and donate any amount of money to help people pay their medical bills.

Elijah Romero-Antoniades *was born in Portland, Oregon, and lives in the Mission District in San Francisco. He is sixteen years old. He has a passion for music, both writing and playing it. He is in a band and also loves skateboarding. Some of his goals for the future are to get a motorcycle and become a tattoo artist.*

HIPSTER

Gentrification has deeply affected both abiotic and biotic factors in the Mission District in San Francisco. Throughout this essay I will be talking about different things that are affected by gentrification and how things react to it. Also, I will talk about how gentrification has affected me and my neighborhood.

The gentrification problem has disrupted the Mission by causing local places to move out and new places to move in. For example, according to a *nytimes.com* article by Carol Pogash, "Luxury condominiums, organic ice cream stores, cafes that serve soy lattes and chocolate shops that offer samples from Ecuador and Madagascar are rapidly replacing 99-cent stores, bodegas, and rent-controlled apartments in the Mission District." These have become popular because they are appealing to the new generation of trendy, rich hipsters. This is a prime example of parasitism because the hipster species can benefit by taking locals' homes and creating other coffee shops that are all generic. In a *citylab.com* article by Tanvi Misra, she writes, "Low- and middle-income families are being priced out of neighborhoods they used to call home." This shows that hipsters are one of the things pushing locals out because they have more money and are at a greater advantage at getting a place where they want to live.

According to a *kqed.org* story by Patricia Yollin, "By 2040, San Francisco County is projected to have a non-Hispanic white majority— jumping from forty-two percent in 2013 to fifty-two percent in twenty-five years. The percentage of Asians is expected to fall from thirty-four percent to twenty-eight percent. The Latino population is forecast to shrink from fifteen percent to twelve percent. The city's dwindling number of African Americans, currently down to six percent, should remain the same." I think gentrification is a result of two key factors: the first being the tech industry and the second being the appealing culture of the Mission. In dealing with gentrification, the Mission has brought a specific type of person urban youth like to call "the hipster." According to Wikipedia, "The hipster subculture is composed of affluent or middle class young Bohemians who reside primarily in gentrifying neighborhoods." In my opinion, these types of people disrupt the ecosystem because they seem to come in large quantities. They raise the rent for locals because they usually got that guap. Also, the Mission is very diverse from our art to our restaurants. No other place in San Francisco is as diverse as the Mission. In my opinion, that's why I think some hipsters are drawn to my community.

"It's a war zone here," says Carol Pogash. This is evidence that the equilibrium is being disrupted. The Mission's equilibrium is being tampered with because gentrification is moving locals out and the ratio of locals to hipsters is highly uneven. I saw this firsthand when my close friend and mentor told me that he would have to get two extra jobs to stay living in the Mission. He thought it was too much and now he's moving back to New York because the price has gone up too much. This proves that the gentrification problem doesn't just affect jobs, but people also have to leave their homes. Not just a couple of things are affected. A whole lot of things get affected when another species invades another's habitat. Some things are left alone, like the art in the Mission. A lot of the alleys are full of graffiti still, so it is good that these things are still going and a part of the equilibrium is being maintained.

In conclusion, the information I have provided should help you think about your community and how your actions or other people's can affect it. Imagine your community and all its rich history being taken over in the span of about five years. How would you react in this type of situation?

Jhoaris Menjivar, *age fourteen, is from San Francisco. After a brief, boring encounter in third grade, Jhoaris discovered a love for soccer in middle school and now plays often. Jhoaris wants to be a pediatrician and work with teens. She hates banana Laffy Taffy, and wants to learn more about her family in El Salvador.*

THE COLORFUL MISSION

Sofia lived in the Mission District in San Francisco until she got evicted from her house on 20th and Capp Streets in December of 2014. She talked to me about her experience: "My mom stayed up late at night to do papers for court. Since my mom worked at home, we weren't only going to lose our house, but also our source of money. The money she had saved up seemed to all go to the rent. There would be times my sister and I wouldn't eat because there was no money for food." This is one example of a story of longtime residents who were evicted because they didn't have enough money to live in the Mission anymore due to gentrification. The Mission is basically changing from one group of specific people to another, when it should be more diverse to benefit both.

Biodiversity can be defined as different types of people that make up an environment. An example of a biodiverse ecosystem is San Francisco. It's biodiverse because this city has different types of people throughout the city. The thing is, different communities have different types of culture and people, especially the Mission District. At one time the Mission District was a place for Hispanic immigrants to start a new life, but now other people are coming to take over their homes. One thing that is continuing to make the Mission not as biodiverse is this racial change. *The San Francisco*

Chronicle writes, "The white population [in the Mission] has increased 22 percent since 2000, and the number of residents who speak a language other than English at home has dropped 13 percent." One ethnic group is replacing another one, so there is less biodiversity in the community.

Another thing that is affecting the biodiversity in the Mission is the change in the economy. For example, *The San Francisco Chronicle* writes, "In 2010, San Francisco's income started to decrease, while the Mission's income start to increase." The Mission now has one of the highest rents in San Francisco. Some one-bedrooms are $3,800 monthly. The biodiversity in the Mission continues to decrease because with the high cost of living, only rich people can afford to live here. This is like predation because people with more money are taking over the homes of poor people. Going back to Sofia's story: she was evicted because her rent was too high, and like her, many Hispanics are leaving the Mission District with hopes of finding cheaper homes somewhere else. Instead of people living together in the same neighborhood, they're just replacing one another.

Right now in the Mission, the lack of biodiversity is a big problem that needs to be solved. Biodiversity would benefit this ecosystem by letting people get to know other people's perspectives and learn from them. For example, there are many homeless people in the streets because the rent is too high, but a lot of people don't know that. People think the homeless are on the street because they did something bad (which is sometimes the case), but really some people are being kicked out of their homes. For example, one day my dad was walking down the street by the exit near 13th and Mission and saw one of his friends. At first, my dad thought he was on streets because he had not gotten over his drug addiction, but the real story was he had been living on the streets because he didn't have a place to live.

Right now, the Mission District is going from one group of specific people to another, but it would be better to be biodiverse and build a stronger community. If there were more biodiversity, that would represent mutualism because everyone would benefit, since everyone would be able to live in the Mission.

The Mission District is just one of many communities that has had problems with biodiversity since the beginning. There used to be mostly Latinos in the Mission, which is why I think they are being so easily replaced. If the neighborhood had been more diverse, then maybe it could have been more stable since the beginning. I believe the lack of biodiversity is a problem that should be solved. Rent should be controlled so that all types of people can live in the Mission. Imagine the Mission of the future, with more biodiversity: people talking to one another, everyone helping each other, fewer homeless people, and cleaner streets. Everyone would be equal and united.

Leonardo David Urrutia *is fourteen years old. He was born in Mexico and has lived in San Francisco most of his life. In his free time, Leo enjoys playing soccer and video games. He also loves to eat. Leo hopes to attend a university to eventually become a coder.*

THE MEXICAN CARTELS

Mexico is like an ecosystem because it contains biotic and abiotic factors that interact with each other. Some of these factors are Mexican citizens and Mexico's borders and territories. Mexico can maintain its equilibrium by creating a community where citizens benefit one another by making and enforcing laws to keep people safe. Currently, Mexico is in a state of disruption because the drug cartels are trying to recruit members and trying to profit off of Mexican citizens by making them sell drugs, getting them addicted to drugs, and involving them in violence. The cartels are negatively affecting Mexican citizens through their use of violence and their involvement with illegal drugs, causing them to suffer.

In the Mexican ecosystem, the cartels are the predators and the prey are the kids they recruit. One way the cartels prey on people is that they force them to sell drugs. They try to make kids join the cartels by going to their school, threatening to hurt their families if they don't join, and by offering them money to help out their family. When the cartels want to recruit new members, they go for the people who fall easily and will easily join the cartel for money or to help their family, and these people are the kids. A way that kids benefit and the cartel also benefits (mutualism) is that the cartel makes easy money off the kids, and the kids also make money so they can help their family. Another reason why the cartel recruits these kids is because they are likely to be loyal. They will never backstab because they are too weak. The cartels are affecting Mexican citizens because they are getting their kids involved in gang activities, and maybe because of their kid, the families will have a chance of getting killed. This is bad because the cartels get families and kids involved with violence and drugs.

Salvador Cabañas was a Paraguayan National soccer team player, and also played for a team in Chile. He then decided to move to Mexico and play for a team called Chiapas F.C. from 2003 to 2006, then he moved to America and lived there from 2007 to 2010. One time after a game he went with his wife to the bar in Mexico City to celebrate. He was shot in the head that night in January 2010. One of the stories that got told was that "La Barbie," an alleged drug dealer for a cartel, came up to him at the bar and said, "I don't like how you played in the game some days ago."

Salvador Cabañas (people said he was drunk) then said, "I don't care, whatever."

Later, he decided to go to the bathroom to do his business and La Barbie's bodyguard, also known as "JJ," followed him. He confronted Cabañas, and said, "I didn't like how you talked to my boss back there." Salvador didn't seem to care, so JJ took out a gun and pointed it to his head.

According to Cabañas, while JJ was pointing the gun to his head he was telling JJ that he wouldn't do it.

JJ then responded, "Yes I will."

Cabañas then said, "Why are you shaking?"

JJ said, "Maybe because you're about to die you're imagining things." Then he pulled the trigger.

Luckily, when he was at the hospital they found out that the bullet went in the side of his head and didn't hit his brain. At the moment, the bullet is still in his head. He can no longer play soccer because of that. In 2012, he tried to make a comeback, but he wasn't in good enough shape to play. He retired in 2014.

By writing about Salvador Cabañas, I'm showing that anyone can get shot for no reason at all in Mexico. There have been a lot of people who have been shot on purpose or by accident in crossfire. This affects Mexican citizens because they don't feel safe, and if they go out in the street they are scared of being shot by cartels. The lives of Mexican citizens are being affected because the cartels are trying to make money off them and they are getting people killed and addicted to drugs.

I care about this topic because I used to be part of the Mexican community, and because my parents are Mexican. This affects me because if I went back to Mexico one day, I would not feel safe and probably wouldn't want to visit again. I'm worried that something might happen to my grandparents because they still live in Mexico, and I don't want a Mexican cartel to manipulate them into doing something bad. I'm bringing this up because I want cartels to stop getting children involved in this drug war they have, and I want more people to know that this is happening in Mexico still.

Marco Marmol *is fifteen years old and is a starting freshman for the John O'Connell basketball team. When he grows up, he'd like to be a professional baseball player (he's a multi-talented athlete!) When not playing sports, he likes to spend his time playing music.*

ART WITHOUT ITS ARTIST

The impact of people getting kicked out of the Mission District in San Francisco is bad. It's leaving people homeless with nowhere to go. They up the price of rent by a ton, and then the people that live in the houses can't pay the rent, so they have to move out to the suburbs in places like the East Bay where they can afford a house. The people that move into the Mission can get mad at the people that have been living there longer than them. It can also happen that the people who have been living there can get mad at the newcomers. The people who have lived there for a long time might have different taste in food and music; the newcomers have money. The Mission is a bright ecosystem with a lot of Latino culture, and it's changing because of gentrification.

According to an article in *The New York Times*, the Mission was once sixty-nine percent Latino. The Mission has lost more than two thousand Latino residents since the year 2000, according to Census Bureau data. The San Francisco tech boom is reshaping communities. Because of the Latinos depopulating the area, some businesses may lose money and run out of business because nobody will be buying anything there. The Latinos are leaving the Mission and this is changing the ecosystem, because the Mission holds a lot of the culture of Latinos and what they do. This culture is an abiotic factor in the ecosystem, because the Latinos have left their mark on the Mission with their food, dances, and music, and now they are leaving because they can't afford the expensive rent. When the people who made the Mission their home leave they take all their food, music, and dances with them. Then they will leave the Mission without its artists; there will be no Mission.

Over the years I've lived here, there have been a lot of huge differences in the Mission. There have been a lot of new structures and buildings. There has also been a lot of reconstruction. When I was seven, Dolores Park was very small. It was mostly all grass, except the park with the sand and the big swings. Now they rebuilt it, and it looks totally all new, with the new tennis and basketball courts and the new park.

I have lived in the Mission since I was six years old. Then I moved to Vacaville for about three years, but I came back and have lived here ever since. When you're in the Mission, you will feel a cool vibe. You will hear a lot of instruments like trumpets, pianos, piccolos, flutes, and drums. You will smell a lot of food too, like nachos and burritos. There are also a lot of different Hispanic cultures, like Salvadoran, Puerto Rican, Nicaraguan, and Guatemalan.

Gentrification honestly doesn't have any impact on me right now, but it could. I live in a nice apartment on 9th Street and there aren't any protests. The rent is not getting raised where I live. It could possibly happen out of nowhere, and I, my mom, and my brother could be homeless if that does happen. I feel like this gentrification situation is like predation. The people that evict the people are the predators and the people who can't pay the rent are the prey.

So gentrification in the Mission is good and bad. The good things about gentrification are the new buildings being built so people don't go homeless, and that Dolores Park got a new design and looks great. The bad things are that people are getting kicked out of their homes, and the impact is very harsh on both the people who move in and the people who move out of the Mission. One way to sum up the impact that it could have on the people that move in is that they might own a building that has a piece of art that has a story, and the newcomer who owns the building might want the wall repainted. Then this would affect the people who live there, and have lived with the art there. The art might have a meaning or something beyond the paint. Then the people who lived with that art there might get overwhelmed or mad that the newcomer got rid of it.

Robert Viray *lives in the Tenderloin neighborhood of San Francisco. His background is Filipino and his parents speak Tagalog. He likes to play basketball, draw, and eat Filipino food his mother makes. There are five children in his family, and Robert is right in the middle. By next year he wants to decide what major he will have in college.*

INDUSTRIAL IMPACT

The Industrial Revolution caused rapid changes in peoples' lives. The Industrial Revolution is characterized as the replacement of hand tools with power-driven machines. It started in England around 1760. There were dramatic changes in social and economic structure. The changes came through inventions, technology, machines, and factory systems. This is still causing harmful changes to our ecosystem.

An ecosystem is defined as "a natural group of interacting elements formed by the interaction of a community of organisms with their environment." Living things that affect an ecosystem are called biotic factors, like animals, plants, and parasites. In addition, there are abiotic factors, which are defined as a nonliving thing that influences an environment, such as mountains and oxygen.

Before the Industrial Revolution, most people worked on farms and produced goods from their home. This production affected the ecosystem less than the changes caused by the Industrial Revolution. As a result of the Industrial Revolution, more and more people went to work in great urban factory centers. The industrial revolution was based on the "multiplier effect." The multiplier effect is the cycle that starts with demand then turns into innovation, which boosts mass production. This leads to more profit and quicker production. As a result, people left farms to work in factories.

The Industrial Revolution was both beneficial and harmful for people. The Industrial Revolution changed the social ecosystem because it affected people, jobs, farming, and production. In this essay, I will illustrate its effect on peoples' lives through three generations of one family. Although the family is fictional, its experiences may have happened to many families. This example is the story of a grandfather, son, and grandson. It will be told in their own voices.

This family's story will show that the Industrial Revolution has had positive and negative effects on the world and how people live. Some of the positives were that it boosted agriculture and sped up manufacturing with inventions and factories. Also, the Industrial Revolution made traveling and transporting a whole lot easier with the invention of the steam train. Inventions like the seed drill really helped speed up farming. Some of the major negative impacts were pollution and child labor. Children and adults both had to work long hours without breaks. This was problematic because pollution leads to a lot of deaths, asthma, and heart disease. Also child labor had a cycle; kids couldn't get out of it and there weren't any opportunities for them to get a good education.

This fictional story explaining the changes caused by the Industrial Revolution begins with the grandfather.

> My name is James Smith, and I was born in 1730. I lived in England on a farm with my wife and children. I spent most of my time on the thirty-acre farm that I inherited from my father. My sons and I farmed on twenty acres growing potatoes, and almost all of the food we ate including vegetables and meat. The meat came from the cows, chickens, and pigs we raised. My sons helped

me work on the land. When the crops were ready we would hire extra help to harvest the crops. My wife and daughters worked inside the house making items such as clothing, baskets, and canned goods. We rarely left the farm because we had to work day to night just to make ends meet. Some of the children married and left home at around sixteen, seventeen, and eighteen years old.

One day my sons and I traveled to town to sell our potatoes, carrots, and apples. We were excited to see a steam engine carrying livestock, passengers, coal, cotton, wool, and grain, and factories being built. This was the beginning of the Industrial Revolution. We talked about how this would make life better, but didn't realize the problems it would bring.

We went home. Then my son decided to leave the farm to work in the factories to make more money. When he left he sent us letters about the working conditions. He saved money, and from time to time would come to visit us back on the farm. We didn't realize he was being exploited. The factory owners were acting in their own best interest but the workers suffered. This is like the tragedy of the commons. Our lifestyles would never be as they are without the changes the Industrial Revolution made.

My name is Jason Smith. I'm James Smith's son. My family and I live in a small cottage on my father's farm. My life was completely changed by the Industrial Revolution. I was planning to work on my father's farm, but my mind quickly changed when my father, brothers, and I went into town to sell some of our crops. When we went into town we saw factories being built and an amazing invention known as the steam engine. When I saw all this, I then decided to work in one of the factories because I thought the work would be easier than working on a farm and I would make money.

It was extremely difficult working in the factory due to the working conditions, working fifteen hours a day with no safety gear, and receiving injuries from the machinery. I would save up money so I could visit my parents and family on the farm. When my son was ten he went to work in a factory because the farm wasn't making enough money to support us.

My name is Jason Smith, Junior. When I was old enough my father told me to work in the factory with him. Even though I was ten, children were allowed to work in factories. And although I didn't go to school my father taught me how to read and write. The working conditions didn't get any better, but work got easier thanks to assembly lines. By the time I started work, they had assembly lines where parts would come passing through. The process would be repeated all day. The farm my grandfather owned and my father grew up on was sold to make space for more factories. We moved closer to factories and rented a small cabin. The Industrial Revolution modernized the country but exploited workers, making life a lot harder. My grandfather, father, and I suffered from these developments. Our suffering was a form of parasitism, because the factory owner and bosses were getting rich at the expense of the workers whose lives had great hardships.

The three generations of this family have shown the impacts of the Industrial Revolution. Moving into an industrialized world, people benefited through mass production, more jobs, and availability of manufactured goods. However, working in the factories also had a negative impact on lives due to unsafe conditions. The Industrial Revolution changed the world.

Nate Mixon *was born in Lancaster, California, but was raised in San Francisco. He enjoys playing sports. By the age of fifteen, he has played basketball, football, soccer, baseball, and lacrosse. He hopes to become a professional basketball or football player. His backup plan is to become a lawyer or a SWAT team member.*

BECOMING LESS OF A SOCIAL COMMUNITY

I like to see the family-owned corner stores and all different races and cultures in San Francisco's Mission District. The Mission is one of many neighborhoods in San Francisco and has historically been known for being mostly Latino. It has parks, murals, and food. The Mission, to me, is a place you can go to meet all different types of people. All of these things are what makes the Mission a community. The Mission is a social community whose equilibrium is being affected by gentrification. Gentrification is what is making the Mission less and less of the community I know.

The Mission is an ecosystem because you have all these different biotic factors, like all the ethnic groups, that can get along and not have too many problems with each other. These groups are always talking and hanging out on the corners, just talking to each other about everyday stuff, like the NBA or NFL games that are going on that day or week. They are also at the parks, just trying to have fun by playing basketball or soccer to make the most of the day. Just like a biological ecosystem, there are also all of these abiotic factors like all of the murals, paintings, and also the buildings and small businesses. These are all things that make the Mission an ecosystem to me because it is so complex. Also, since the Mission has started to change, there are fewer people just hanging out and talking and a lot of the small businesses have closed over the years.

In addition to the small businesses getting closed and bought out, the number of evictions has been growing very drastically. According to *The San Francisco Chronicle*, "Since March, the number of eviction notices filed per month with the San Francisco Rent Board is up 32 percent compared with the previous three years' average, while owner move-in evictions are up 131 percent." This shows that the neighborhood is probably going to be getting more revenue, but it's not going to be as diverse or be as much of a community as it used to be. A lot of the evictions are being caused by rising rent. Also the number of evictions is being affected by the carrying capacity of the Mission. There are so many more people coming into the Mission that the housing market is at its carrying capacity. There is no more housing for people, but more people keep coming and the people that get pushed and forced out are just becoming homeless or moving away. It is also happening to small and family-owned businesses. All of these small family-owned businesses are getting bought out and moved to create more condos and living space that is getting filled just as fast, if not faster, than the spaces are getting built, and as this is happening, the Mission is becoming less of a community.

"Rents are through the roof, and investors are gobbling up multifamily buildings at unheard-of prices—apartment complexes with more than 10 units sold for an average of $357,000 per unit during the first half of 2015, nearly double the $180,000-per-unit values seen in 2011. Investors paying that much money for apartment buildings have a huge incentive to evict tenants paying below-market rents." This quote from *The San Francisco Chronicle* shows that the Mission living and working community is affected negatively by the rise in rent. All of the big business workers are the ones causing the rent increase, and the rise in rent is causing a lot of the working class to move or get kicked out. Although this negatively affects the Mission community, it does not negatively affect the landlords because they are getting more and more money from the companies and the workers that want to move here to the Mission.

Predation means basically that the big fish will always eat the little fish. The big fish are all of the big brand tech companies and their workers, and the small fish are the small businesses and the people that don't have enough money to pay the rising rent. The big companies are gaining business and their workers are getting a good place to stay, but all of the small businesses and lower or middle-class workers are getting nothing. They are just getting kicked out of the place in which they are living and also getting displaced and fired from their jobs, which in the long term may cause them to be homeless or desperate.

San Francisco is where I have lived my whole life. The Mission is where most of my time was spent. The Mission District matters to me because it's right around where I've grown up and spent a lot of time as a kid, and now it is also where I still spend most of my time. It's the most diverse place that I have been and it has always been fun to go down there and just hear everyone talk and see all of the paintings. Ever since I could, I would try to come down here to the Mission and hang out with friends and just take in all of the elements.

The Mission is a social community whose ecosystem's equilibrium is being affected by gentrification and in the long term, the effects of gentrification are turning it into less of a social community. This is a bad thing, because when you have less of a community it becomes more likely that there will be more crimes and no one will feel safe if that happens. So after reading this and finding out what I know about the Mission and what I was able to find out from my research about these types of evictions and rising rent prices, what do you think the Mission will look like five, ten, or maybe even twenty years from now?

Max Sedakou *is from Belarus. In spite of getting several injuries from the sport, he loves skateboarding. He also likes biking, hanging out with his friends, and listening to rap music. One day he hopes to move back to Belarus and become a mechanic or a pro skater. One of Max's favorite things is his grandma's homemade cherry juice on a hot day.*

375 TO 415

Hello. My name is Max and I am sixteen years old. I'm from a little country called Belarus, between Russia and Poland. I lived there for twelve-and-a-half years before I came to the United States in 2011. I live here with my mom. My granny stayed in my country with two of my favorite cats. Belarus and the U.S. are both ecosystems. However, they are different in several ways. They are both ecosystems because they have biotic factors, such as people and animals, and abiotic factors, such as land and money, but nevertheless they are different.

The reason I came here is because I was about to get sent into the army. In Belarus, every boy who is age eighteen must to go to the army in case of a war. They learn how to use guns, clean them, fix them, and they learn how to fight. My family changed our ecosystem because they didn't want me to go to the military. They decided that I should go to a different country to explore new things.

The army and Belarus are both like ecosystems. Sometimes in Belarus a little boy might get bullied by his classmates before he goes into the military. This is an example of predation. Once he goes to the military he learns how to fight, how to use guns, clean guns, and fix guns. He learns how to survive. So once he comes out he could actually come back for payback for his old bullies. Now the little boy is the predator and the bullies are the prey.

I'm lucky that the government doesn't make all the boys join the army here. The government here also doesn't make bus drivers do a lot of medical, mechanical, or mathematical tests. In my country, every bus driver must pass about ten medical tests, and they must pass a drug test every month. From what I know, the bus drivers don't have to do all those things here in America. The ecosystems are different, here and there.

Another difference is that there are no homeless people in my country. If you failed high school there, you can still go to a continuation school, then college. There are big differences between schools here and in Belarus. The schools there are way harder than they are here, meaning that finishing high school in Belarus means a lot. My friends told me that. In Belarus, in ninth grade, kids learn college-level subjects. For example, I had biology in sixth grade. And did you know that in Belarus, kids in fourth through eleventh grades stay in the same school?

Everyone has a job in my home country. If someone doesn't want to work, they must pay a fine. If they don't pay the fine, they have to work for no pay or they will go to jail. People from my country care about the country. People don't do drugs on the street. Everyone takes care of each other. This is called altruism. Here in America, if someone failes high school they can't go on to college, and they might become homeless and stay broke.

Between my country and America there are a lot of different group behaviors. I really miss my country, even though I just went there over the New Year break. I miss my grandma, my cats, my classmates, and how pretty it is there. But I'd rather stay here in America because there are a lot more things to do, like skate, go on social media, and make friends. So right now I live in San Francisco. I like it here because there are a lot of skaters and skate spots, but I want to go visit Belarus soon.

Luna Martinez-Trejo *has been writing poetry since seventh grade. To Luna, poetry is an important form of expression, and she uses it to stay in touch with things that are meaningful to her. She loves food, the color red, the heat, the relaxed pace of summer, hot sauce and movies that scare her or make her laugh. It is important to Luna to remember that pain never lasts.*

ESTA ES MI RAÍZ Y DE AQUÍ NO ME VOY A IR

Just a walk in tha streets
will bring so much emotion,
like seeing "Lilo & Stitch" on a bright screen TV
You'll feel tha vibe, the trust
the neighbors got in us,
the happiness, the love
that will always be above.
Above the importance of money.
It's more important getting family reunions,
drinking some tea with a lot of honey,
that sweet taste
is sweet as seeing a child smile on a sunny day.
The type of family we have,
the bond u will feel and see,
the vibes we give out,
is something you will rip out.
Takin' away our house is like popping the vessels in a heart.

It will soon be ice cold,

colder than any soul in this world.

So please don't take my home,

please let us be.

Taking this away is cutting the roots of the tree.

The branches, the way they blossom,

is our culture that gives the tree a beautiful look.

Our culture will fall slowly.

Dejanos vivir porque me quitas

mi casa a mi

qué sientes con destruir

nuestra cultura y pisas en mi raises

no puedo caminar

sin que se me queden viendo

como si yo no pertenecería aquí

esta es mi raíz y de aquí no me voy a ir.

Gentrification in San Francisco's Mission District is a personal topic for me because it's a really current problem that my community is dealing with. I've seen so many articles about this issue, but they never really tell you about the emotions that go on throughout the process of gentrification, and the harm they do to individuals and families. This is also related to the topic of housing policy because that's a big part of gentrification in the Mission.

On this website I found, *sf.curbed.com*, Deb Follingstad described her experience with skyrocketing rent resulting from gentrification. The article says that on March 1, Deb Follingstad and her boyfriend were renting an apartment in Bernal Heights, paying $2,145 for a two-bedroom apartment. The next day, March 2, she found herself reading a notice informing her that her rent would soon jump to $8,900 per month. Deb Follingstad said, "I understand that a rent-controlled apartment is a ticking bomb." She meant that any minute this "rent control" can explode to an extremely high rent. She made a post on Facebook that went viral. She had posted a scan from the legal notice, which was shared more than 2,600 times.

I attended a workshop here at John O'Connell High School where they discussed the topic of gentrification. In this workshop, I learned how high the eviction rate was due to The Ellis Act (which gives the landlord the right to evict tenants in order to "go out of business"). We also discussed which area was being targeted the most, and it was the Mission and the areas around it. I learned that sixty-nine percent of the evictions occurred within four blocks of known shuttle stops—tech buses for Yahoo, Google, Facebook, Apple, and Zynga. From this I learned that one of the major reasons why "techies" move to our neighborhood is because all the tech busses are near here.

I wrote a survey because of the life-affecting impact that the process of gentrification had on my family. In the survey I asked my family members to answer questions about their experiences. My intention was to be able to share the emotions that we feel when we experience gentrification. I really wanted you, the reader, to understand the inside of what they feel and what their experience was.

Most of the people I surveyed used to live here in the Mission or have lived in the Mission for ten to thirty-five years. Most of the people that I interviewed were in my family, and lived with me in a house at 20th and Alabama Streets. My grandma, Maria Espadas, lived in that house for twenty-five years. We had about thirteen people living in that seven-room house. My grandma paid a reasonable rent of $1,000, and my uncle was the landlord so he tried to keep it stable for us. We ended up being evicted from the house because the rents all around us just started shooting up out of nowhere and he couldn't afford to keep it cheap for us anymore. We no longer live there.

To the fifteen surveys that I did, I received the same kinds of responses. People said they felt very angry and depressed about what's going on. They said they felt uncomfortable in their own neighborhoods. They said that when they walk through the streets of the Mission, to a park or a corner store, they feel out of place. Some families complained about how they tried to enroll their kids in Spanish immersion programs to support their children, but they were filled up with non-Hispanics. The sense of safety in the neighborhood has been disrupted by many things, such as frequent fires on Mission properties. The impact has hit harder than a tsunami during the last three years. The displacement has been more than physical, it's also hugely emotional and psychological.

This compares to the biological term the tragedy of the commons, which is a situation where individuals act in their own best interest but at the big cost of the group's success and everyone involved is eventually hurt. Our limited resource in this situation is the housing in the area. There's not enough, and when the newcomers or developers act only in their own best interest, we all lose; the bright, cultural Mission that you once saw in pictures or you have in your memory has been made dull.

As you can see, this problem runs way deeper than money. This affects people emotionally. This is how it has affected me. It has made me feel isolated from my family. The vibes we had in that house on 20th and Alabama were great! We had love and trust. And our culture is all about family.

Something I have noticed is that the tech culture's focus seems to be on money and having a very "independent" life style—for instance, living in a large house with just three people. This stereotypical lifestyle might make them look happy, and some people might be happy with being alone. But I really doubt it because if they were, they wouldn't be coming to the Mission, seeking the vibes and mystique of our heartwarming culture.

IF WE COULD
INTERVIEW A WHALE

How we impact the world around us

Zakaria Kerrar *is in his freshman year at John O'Connell High School. He spends his free time longboarding in the city and taking long-distance rides to as far away as Stinson Beach. He intends to become a famous drummer in the not-so-distant future.*

OCEANS NEVER LIE

Global warming, climate change, and ocean ecology are very complicated systems. Many factors are influencing the health of our planet. Whales are affected by rising sea levels, pollution, commercial fishing, rising temperatures, and chemistry changes. To learn more about how the oceans are changing we can focus on whales because they migrate, and some whales travel between the tropics and the Arctic. If only we could interview a whale, we could learn a lot.

There are many ways that humanity influences the ecological health of the world's oceans. Two big ways that humans have affected the oceans are whaling and burning fossil fuels.

When people use fossil fuels, it contributes to global warming, which in turn melts Arctic ice and causes sea levels to rise and disrupt the food chain. If ice melts in the Arctic, whales' food sources are diminished. Also, there are places in the ocean where trash collects in huge trash islands. The trash all flows to one part in the center of each ocean current and creates what scientists call a trash gyre. Another major impact that humans impose on the world's oceans is commercial fishing and whaling.

The oceans are all interconnected. In fact, you could say there is only one ocean on the planet with many currents and diverse life forms. Some whales, such as grey, humpback, and blue whales, actually pass by the San Francisco Bay on their way to the Arctic in some seasons, and the tropics or Baja California in other seasons. Some sea animals travel the thousands of miles connecting distant regions of the earth. Two good examples that pass by the Golden Gate every year are whales and salmon.

Everyone is talking about how the climate is changing and the earth is warming up, but I want to know how these forces and changes are affecting the oceans of the world. While I was researching global warming in the Arctic, I discovered a major glacier in Greenland that shares my name. I don't know why they call Greenland green, because as of now it is almost all ice. But signs show that the ice is in danger of melting! In fact, a gigantic glacier named none other than Zachariae Isstrøm is melting faster and faster. Some scientists think that the Zachariae glacier could raise sea levels about half a meter globally! That is some serious water! I want to travel to Greenland and check out the Zachariae glacier before it melts completely, because in my imagination it is named after me, even though it was named before I was born. If I were to travel to the glacier, I would most likely travel by snowmobile or dogsled. I would prefer to take a dog sled. Huskies are amazing dogs, and they have so many eye colors. I wonder how many dogs it would take to get to the glacier safely. I would guess eleven dogs or more to travel.

The fishing industry, the oil industry, and the shipping industry create challenges for whales and other sea life—like in the reality TV show *Whale Wars*. On this show, environmental activists and organizations like the Sea Shepherd Conservation Society take matters into their own hands and try to slow down the destruction.

The Japanese ships say they are commercial ships and are just trying to film whales, but in reality they are fishing for the whales. Sea Shepherd attempts to stop whaling practices, and they attack whaling ships with smoke bombs. Once Sea Shepherd got a new ship, brand new, with guns that shoot smoke bombs, not actual bullets. They surrounded three ships—I don't know how—with little ships. These people risk their lives to disturb the whaling industry. They want to stop people from endangering endangered species.

The good news amid all of these challenges to our world's oceans is that we can do a lot to shift the ways in which we interact with the sea. The ocean is like outer space; it is almost impossible to understand how vast and deep the sea is. I can't believe that people have finally become a threat to even this immense ecosystem. It is time to stop taking so much life out of the ocean and agree on polluting less. The ocean is like a history book. All of the climate change is recorded in the sea temperature and sea level, so it is insightful to study the ocean and see how the world as a whole is functioning.

The ocean is like a mirror for humanity. The ocean is another dimension; when people go fishing or drilling for oil they encounter many challenges and realize that they are not in control of everything. The oceans share borders with almost every nation on earth. It is a universal connection that we all have, so we should take better care of our relationship to the sea.

Addis C. Riggen *was born in Denver, Colorado, but now resides in San Francisco, California. During her downtime, Addis enjoys playing role-playing games. Her cat, cleverly named Kit-Kat, is a four-year-old Calico who is extremely sweet. Addis also enjoys reading horror stories, particularly* Scary Stories to Tell in the Dark.

MOVING IN, MOVING OUT: GENTRIFICATION IN SAN FRANCISCO

Living in San Francisco with my family, I've noticed many outcomes of gentrification in my everyday life. The prices of newly built apartments are far too high for most people to even think about living there. One apartment building I live near was listed for $4,000 per month. This excludes people who do not make very much money, and is a clear indicator of gentrification. I have found that when a neighborhood is being gentrified, it is very similar to ecological interactions found in nature, such as parasitism, predation, and the idea of an invasive species. These types of interactions often end up disrupting the balance of the neighborhood through real estate and schools.

In San Francisco's gentrified ecosystem, there are many biotic and abiotic factors that interact together. A biotic factor is a living thing that affects an ecosystem. Generally these are plants, animals, and microorganisms. An abiotic factor is a nonliving thing that affects an ecosystem. These include rocks, water, and oxygen. In San Francisco's gentrified ecosystem, the biotic factors are low-income workers, poorer families, students who go to public schools, richer families, and students in charter or private schools. Some abiotic factors in San Francisco's gentrified ecosystem are tech companies, public schools, incomes, rental prices for housing, and charter or private schools.

Some of the ways biotic and abiotic factors of this ecosystem interact with each other are gentrification itself, which is when highly paid workers move into the poorer, but oftentimes well-established neighborhoods, forcing the prices to rise and the low-income families to move. This resembles parasitism in the way that the high-income workers (the "parasites" in this instance) are benefiting at the expense of the low-income workers (the "hosts"). It is this type of interaction that has a major effect on the ecosystem.

Gentrification is caused when wealthier residents move into poorer neighborhoods and cause rent prices to spike in a community. After the wealthy people move in, the poor are forced out due to rent becoming too high. This disrupts the balance in the community, as families who may have lived in the same place for a long time are suddenly forced to move somewhere else. A wealthy newcomer's sudden move into a poorer neighborhood is an example of how an invasive species affects an ecosystem in nature. The wealthy people are the ones "invading" the poor neighborhoods and this trend has increased at an alarming rate, just like how an invasive species spreads rapidly.

One of the most prominent effects of gentrification is that the landlords will raise the price of rent in their buildings because of rich people moving in. Here in San Francisco, communities that have been established over generations are being uprooted by tech companies and their highly paid workers. An article from *The Huffington Post* states, "While the city's median family income of about $103,000 is well above what's needed to afford to live in San Francisco, high rents have the effect of pricing people on the lower side of the economic spectrum out of the market almost entirely." Because our communities are being gentrified, the rates for rent all over are skyrocketing. Landlords are raising the rates to accommodate the richer families moving in.

Along with its effects on housing and rent, gentrification also has effects on public schools, especially on their performance and enrollment rates.

Gentrification in public schools causes tension, and eventually protests will start up if they have not already. The public schools also have lower enrollment rates due to the higher-income families sending their children to private or charter schools instead of the public schools. An article from *The Atlantic* magazine states, "Because newcomers tend to send their kids outside of the local system, often to private or charter schools, gentrification tends to have a neutral or even negative effect on neighborhood schools, at least in the short term." The same article in *The Atlantic* talks about how public schools are trying to adapt to the newly gentrified neighborhoods around them: "Sarah Garland of The Hechinger Report suggests that the best way to have gentrification help local schools may be to invest in more and larger magnet schools and bring more diverse students into gifted-and-talented programs. One county in Florida has had success doing just that, finding remarkably talented kids in poor neighborhoods that school administrators had, up until now, neglected." This is clear evidence that if the public schools manage to adapt and keep up with their gentrified neighborhood, then the school will have a much larger number of children going to the school. Those schools have achieved this by offering more gifted programs and better academic choices.

The evidence from my findings shows that gentrification shares many similarities with an ecosystem in the wild. The ecosystem I have decided to work with is the city of San Francisco. Within this ecosystem, the tech company workers behave like an invasive species when they move into poor neighborhoods and make the rent prices climb at the expense of the poorer families gradually becoming displaced. Gentrification is a severe problem in San Francisco that will eventually widen the gap between the rich and poor population, based on my experience with the $4,000 rent for a two-bed, one-bath apartment, to the many online articles talking about the effects of gentrification on schools and neighborhoods alike. If this doesn't stop soon, many people in my generation may not be able to live in San Francisco or go to college here because the price for rent is too high in too many places.

Arcel Navarro *leads a life that revolves around his love of soccer. As a starting forward on Sport Team Santa Clara, he travels to the South Bay every weekend for practice and matches. A San Francisco native, Arcel grew up in the Mission District. Arcel likes school, and knows there are many more years of it ahead before he will achieve his goal of becoming a pediatrician.*

YES, CARS LEAVE FOOTPRINTS

When I started traveling to Santa Clara with my soccer team twice a week, I noticed how many cars were on the freeway. I began to think about how much gas was being used and the impact on the ecosystem, in my case, San Francisco's Mission District. While I see a lot of cars in my neighborhood, I didn't realize how many were on the highways around the Bay Area. So how does the carbon footprint vary among the different neighborhoods in the Bay Area and why?

I found research from the CoolClimate Network at the University of California, Berkeley, that estimated carbon consumption by neighborhoods throughout the Bay Area. I saw that my neighborhood, the Mission District, has a low carbon footprint compared to other neighborhoods in San Francisco and in other cities.

For example, the Mission District's average annual household carbon footprint is 38.1 metric tons of CO_2, but the Sunset District's carbon footprint was 46.4. The research showed the main reason the Mission District's carbon footprint is low is density. With more people sharing space, they also share resources like lighting and heating, and services like police, fire departments, and schools. Less dense neighborhoods like the Sunset District and the suburbs of San Francisco, such as Santa Clara and San Mateo, have a higher carbon footprint.

I also noticed in the research that transportation is the factor that contributes most to carbon consumption. That made me think of my soccer trips to Santa Clara. The people in any neighborhood or city (pedestrians, bicyclists, and drivers) are the biotic factors in the ecosystem. The cars, busses, and bicycles are among the abiotic factors in the ecosystem. Just like an ecosystem is more stable when there is a greater diversity of species, the Mission can benefit from a diversity in transportation choices. Even with fewer cars, people can still go where they need to go by public transportation, bicycle, or by foot.

A *Berkeley News* article about the CoolClimate Network project said there is a "wide variation in size and composition of household carbon footprints." In my ecosystem, the Mission District, it is fairly easy to avoid driving cars because the area is compact, and there are a lot of public transportation options and bike lanes. In suburban areas, people may need to drive more because of the distances and lack of good public transportation.

Carbon consumption is important because it can cause air pollution and can affect climate change. A *San Francisco Chronicle* article about the Cool-Climate Network research found that suburban neighborhoods have a larger carbon footprint "because of factors such as having larger homes and household sizes, a larger number of cars owned in general, and longer average driving distance." Meanwhile, dense neighborhoods like the Mission District "tended to have lower carbon footprints because of their denser, affluent housing, accessibility to public transportation, and fewer costs associated with household heating and cooling and goods to fill the home." The article said the survey shows that "people who live in suburban areas need to take a deeper look at the way they live." For example, they should look for ways to drive less or carpool. When they have to drive, they should choose a car that gets more miles to the gallon.

No matter where someone lives, people who drive more increase their carbon footprint. Their air pollution affects everyone, even those who drive cars less or not at all. This is an example of the tragedy of the commons, where individuals act in their own best interest but at the cost of the group's success. By driving less, these people can improve the air quality for themselves and their neighbors.

Even though the Mission District has a low carbon footprint compared to other neighborhoods in the Bay Area, there is room to improve, especially in the area of transportation. The city government should improve the streets to make them safer for pedestrians and bicyclists. For example, there should be better lighting on sidewalks and streets, more speed bumps on side streets, more countdown signals at intersections, more enforcement of speed limits, and wider sidewalks. To improve bicycle safety, I suggest more bike lanes, more training of drivers to respect bicyclists and training of bicyclists to follow the rules, a Bike Share station in the Mission, and other services to make biking safer and a better alternative to driving a car. Plus, the city government should provide more and better public transportation by upgrading the system.

Providing more and better ways for people to move around without a car is an example of mutualism, in that it is a win-win situation for everyone. Even drivers benefit from cleaner air and less traffic. On the other hand, building more roads and parking lots within an ecosystem like the Mission District is an example of parasitism, in that it encourages more driving and benefits drivers, but this causes more air pollution, which harms everyone.

Even suburban areas can lower their carbon footprint by providing more alternatives to driving cars. My research shows, for example, that the city of Santa Clara in the South Bay has a very high average annual household carbon footprint of 52.3 metric tons, with most of it coming from transportation. The government could do more to encourage the use of public transportation and bicycling.

My carbon footprint research shows not only where a carbon footprint is high or low, but gives us clues about what can be done to lower the footprint in neighborhoods across the Bay Area and across the country. A lower carbon footprint can mean cleaner air and can reduce the effects of climate change.

Diego Lopez *is sixteen years old and very talented at video games involving hand-to-hand combat. He likes spaghetti and listens to many different kinds of music. Diego hopes to get better at singing and playing video games. And look out: he throws a mean snowball!*

THE CHANGE OF THE WORLD

Our world would be completely different if the Industrial Revolution never happened. In the Industrial Revolution, people created new inventions to make resources, like food and clothes, faster. After more time, more inventions were made, and eventually this led to our day and age where everything we want comes to us easily. This made us accustomed to having everything made for us, to the point where we can't make most things ourselves, like food and clothes. Although it led to many good things, like advances in technology, getting everything we need whenever we want, and people being able to have their own businesses, it also led to a large amount of pollution, to people getting hurt and killed by the terrible conditions of factories, and to people being exploited for profit.

Before the Industrial Revolution, in the farmer days, the ecosystem consisted of farmers interacting with other farmers, people, and animals. People were both producers and consumers because they would farm food for themselves and for other people who wanted to trade for things they needed, like tools for their crops and clothes. Neighbors trading with each other developed a mutual relationship, where they both depended on and helped each other. Because they depended on each other if a tragedy of the commons happened when someone took more resources than the person needed and messed up the crops, it would hurt everyone. For example, if farmers had livestock, like cows, and someone added more cows to have more resources, like milk and cow meat, the larger number of cows would eat more of the common resource (grass). Then the grass wouldn't be able grow as fast as the cows need it, so they would starve from lack of food. This would hurt the people with the livestock and their neighbors depending on their cows. This shows how mutualism was a common relationship between people where they depend on each other to make the resources they need. But that also led to people having to keep their crops healthy for themselves and for their neighbors, who also depended on their resources.

During the Industrial Revolution, new inventions changed the old ecosystem because the people weren't getting their resources from farmers anymore. Instead they got them from factories. People had to leave the farmer life and move to cities to work in factories because the new inventions were faster and better quality. The factories produced so much for less money that they could afford to sell their products for a cheaper price. That led to people getting the products they needed from the factories instead of from the farmers. In this new ecosystem of factories being the main producers, there was a lot of parasitism. People who owned the factories would benefit from the workers by noticing the fact that their workers were desperate to keep their jobs to get money to feed themselves and probably a family. They couldn't afford to lose their jobs since there was a limited amount of jobs and many unemployed people were ready to take any job. The factory owners didn't want to waste their money to make their factories safer for people, so the workers would work in terrible conditions where they breathed in a lot of bad gasses and the machines would take fingers, limbs, or even their lives because the factory owners would lose money if they made it safer for the workers. They were slowly dying, but the factory owners wouldn't do anything because of their greed. They would even use children because they were easier to manipulate and factories would have to pay them little to none. People's greed can lead to the suffering of other people.

These new inventions led to the ecosystem we know today, where we get everything we want faster and better. Factories are now farther away from us, so it's easier to ignore the same parasitism that happened in the Industrial Revolution. For example, in the chocolate industry, they need cocoa and they get this from Africa, where a lot of people, including children, are working with dangerous tools and getting paid little to none. The parasitism is happening between the companies and the workers, and it's the companies who are being the parasites. At the same time, since the Industrial Revolution, we have advanced a lot in the medical area. We started to live longer, we get everything we want whenever we want, and we have more of some resources, like clothes, than we could ever need. We also have mutualism in our present day, but it isn't as important as it was before the Industrial Revolution in the farmer days because they couldn't make everything themselves or go to a store to get the things they needed, so they would depend on others to make the resources they couldn't.

I think that in our day and age, factory owners should think about their workers more and give them better conditions so there can be fewer accidents. They should also pay them more so they can live a better life. From this they would lose their parasitic relationship and gain a mutualistic relationship because the workers would make their products and they'll make money, and workers will get paid more to live the life they want. The Industrial Revolution gave us new inventions so we could get things faster and better, but the inventions were also used to fuel some people's greed. Without the Industrial Revolution, we would still be living in the farmer days, tending to the crops and livestock without the technology we have today. So what do you readers think? Would you prefer for the Industrial Revolution to have happened or not?

Kasandra Lara Montes *is a sophomore at John O'Connell High School. She lives near downtown San Francisco with her parents and older brother. She enjoys walks around town and reading, especially books with humor. She hates Internet pop-up ads. She's been told that she's good at math and science, but she's not so sure. (The editors of this book are quite sure she is.)*

WHY SHOULD WE CARE FOR THE BAY DELTA?

The San Francisco Bay Delta is a very important ecosystem for California because it supplies us with fresh water for our daily use. The Sacramento and San Joaquin rivers pour fresh water into the Delta, where it mixes with salt water. This process creates a unique ecosystem, providing a habitat for hundreds of species. For over a thousand years, this ecosystem was in balance until humans came along.

Humans have been taking more fresh water than necessary, which causes the Delta to slowly lose fresh water, and the native fisheries to lose business. In order to save this ecosystem, we must come up with a plan to ensure that both humans and fish get the water they need. The Bay Area has been my home for sixteen years, and I wouldn't want to move simply because I don't have any water to use. I am one of the nearly seven million people who live around the San Francisco Bay and want to ensure that the Delta is healthy and in equilibrium.

An ecosystem is a natural system where biotic factors (living things) interact with abiotic factors (nonliving things) in their environment. The Bay Delta is a water and wetland ecosystem where fresh water and salt water meet. This ecosystem consists of both factors; the biotic factors include endangered species, such as the Delta smelt and humans, and the abiotic factors include fresh water, salt water, and the pumps that divert the fresh water south for farmers. Right now, this ecosystem exhibits ecological competition. Humans and the fish both need the same limited resource: fresh water. Since the water is running low, the farmers are losing their crops. And the Delta smelt, for example, are dying because the water is not fresh. Sadly, this causes harm to everything in the ecosystem.

Droughts are one of the main reasons the Delta is losing fresh water. According to UGSG (Urban Geography Specialty Group), two scientists, Michael Dettinger and Michael Anderson, have investigated the water reservoirs' storage and noticed that the snowpacks are dangerously low. "Thus in 2015, California's major reservoirs—which are important tools to manage water supply through drought conditions—did not receive the snowpack runoff necessary to refill them after three years of drought," they said. Snowpacks are the key source for fresh water, but with a long-term drought, there won't be any more snowpacks for a very long time.

The increase in human population is also another main reason the Delta is losing its fresh water. According to *The San Francisco Chronicle*, the aquaduct that moves water through the Sacramento-San Joaquin River Delta "to cities and farms in the arid south, was completed by the 1970s when California had just 16 million residents, compared with the 40 million who live there." Can you imagine how much more fresh water we would need with this amount of people? The Delta will be gone if more humans are born at this rate.

The Bay Delta's endangered species have experienced major problems that can leave them extinct. One example of a species facing such problems is the Delta smelt, a type of fish which has been listed as endangered since 1993. According to *National Geographic*, these fish are close to extinction due to lack of fresh water. "Only six delta smelt—the lowest number ever found—were netted in a survey by state biologists," according to the magazine. Six delta smelt in the sample size is the lowest number they have ever seen. On the other side, farmers are losing their businesses because they have no water from the Delta to grow their crops. According to *The New York Times* the Westland Water District, which is on the west side of the San Joaquin Valley, has been giving tons of water from Delta aquaducts to the farmers. But because of the current California drought, the water has become little drops. If we keep losing our fresh water, then our environment will be as dry as a desert, or worse.

With these major problems occurring, there have been debates about what to do with the Bay Delta. Governor Jerry Brown has proposed the Bay Delta Conservation Plan. One element of this plan is to dig massive tunnels to secure the fresh water from the rivers, bypassing the Delta. According to a group called Restore the Delta, government officials claim that the project will cost $17 billion for construction. These tunnels will grab two-thirds of the Sacramento River's water, which is fresh water from the snowpack. However, adding these tunnels could make the problems worse. If we took all of the fresh water from the Delta, it could have more impact on hundreds of species of wildlife in and around the Delta and coastal fisheries. The habitat would be gone, the economy would be bankrupt, and the water could be contaminated from the tunnels.

Luckily, there is another solution that is less damaging than the conservation plan. The organization, Restore the Delta, states that "all the Delta needs is restored water flows and levee upgrades." Levees protect the water supply and its quality. To fix the levees, it would cost $2 billion to $4 billion, according to the Delta Protection Commission. Even though we have a solution for the levees, they still need a solution for water flow. Water flows are good for fisheries, such as salmon fisheries, since salmon help to keep pollutants away from the Delta. "Extra water can be exported through the existing pumps if state-of-the-art fish screens are installed by the water takers as promised," Restore the Delta says. Fish screens are designed to protect fish from swimming to rivers where the water is taken out for humans. These screens can protect the fish from being killed, while making sure we receive extra water. This plan is safe and costs less!

Overall, the Bay Delta is in serious danger and humans are the only ones that can save it. By voting for the Bay Delta estuary plan, we are saving ourselves from paying billions of dollars and using dirty water for daily use. This plan can preserve fresh water and keep the fish swimming. However, if you don't like this idea, then there is one simple thing you can do: save water! The tragedy of the commons has been going on for a long time because we can't stop taking more resources than necessary. Everyone around us is being affected by this dilemma, so we must change our actions for the Bay Delta ecosystem!

Samuel Esquivel *was born in St. Paul, Minnesota, but lived in Florida and Colorado before settling in San Francisco in 2015, where he lives with his father. He is passionate about reading and watching anime, and has a wry sense of humor. In the future, he plans to have a career in technology.*

HUMAN EFFECTS

Humans have been messing up things since the birth of well… humankind. We've committed mass murders, mass scale destruction of cities, forests, and ecosystems, and now we're changing our own planet's global temperature. Like, I understand changing the temperature in your home, a small region even, but the whole dang world?! How in the bloody tarts did we manage to do that? Well, that's what I'm here to explain: how humans have caused global temperature to increase.

Now, let's get to the boring, yet important, facts. I know that climate and temperature changes are caused by human use of fossil fuels, according to major and credible organizations who have shared the recorded information on the release of mass amounts of harmful CO_2 (carbon dioxide). Fossil fuels include coal, natural gas, and oil that we use in everyday things like cars, buses, kitchens, electricity, and lighting in our homes and workplaces. Fossil fuels release gas when they are burned. For an example, when a car burns gasoline, it releases CO_2 into the air. In the United States, 210 million people are licensed drivers, which results in a ridiculous amount of carbon emissions.

Our ecosystem, the world, is a system of interacting members. It is being hurt by the humans, one of the biotic factors. A biotic factor is a living organism, just like an abiotic factor is a nonliving organism, like a rock.

Humans are the ones doing the harming, while still taking more resources from our environment; we could fall under the category of a parasite: an organism which harms the host while taking resources from it. We are Earth's parasite.

It is estimated that by 2085, whole countries will be submerged due to glaciers, polar ice caps, and Greenland's icebergs melting. According to an article in *The Contra Costa Times*, 2015 was the hottest year on record, beating the previous year, 2014, by an astounding 0.24 degrees Fahrenheit. This global temperature rise was recorded by the National Oceanic and Atmospheric Administration (NOAA) and the National Aeronautics and Space Administration (NASA). Currently, this temperature rate seems to keep increasing yearly, with no evidence of stopping.

While it doesn't seem like too much of a difference, all of the major climate problems you always hear about are occurring because of the temperature rise. These CO_2-caused increases affect important places like the polar ice caps. A huge amount of ice, when melted, could cause global flooding. Evidence of this has already popped up in the Maldives, a small island chain near India. As the glaciers disappear, so does this country. This is one of many countries estimated to disappear by 2085. Other countries such as Tuvalu, home to 9,000 people, are predicted to disappear in at least thirty years. A few countries in danger include the Philippines, Egypt, Barbados, and Papua New Guinea.

A new and heated topic is the TPP, otherwise known as the Trans-Pacific Partnership. If adopted, this partnership would protect fossil mining and usage companies, giving them the ability to sue countries for setting limited mining production. This proses a major problem to our already diseased planet, because, without restrictions, these companies could mine and release even greater amounts of CO_2. This would further the global temperature increase problem, and eventually, the temperature as well.

We could not talk about how this would reduce tariffs (taxes) on the companies because it's irrelevant. WRONG! Of course it's freaking important! The less corporations pay in taxes, the more money they have to continue mining, polluting, and paying off anybody who gets in their way.

. Another major problem is large acts of deforestation. Large forests hold big amounts of CO_2 captured by trees. When trees decay and die, the CO_2 is released into the air. An example of this is the Amazon rainforest, which holds about half of the world's captured CO_2. As more CO_2 is released and more trees are destroyed, fewer trees are around to capture the never-ending amounts of growing CO_2. It is estimated that up to eighty percent of the world's trees have already been cut down. Imagine how much CO_2 has been released if an average tree in a rainforest holds up to hundreds of years worth of CO_2.

Let's talk about something that most causes of pollution can fall under: tragedy of the commons. It is when humans act in their own benefit, which ultimately winds up hurting everyone. An example of tragedy of the commons is unregulated over-logging. In the Amazon, every day, hundreds and probably thousands of acres are cut down illegally for industries establishing new cattle farms, factories, and houses. About 224,000 square miles have already been lost in the Amazon since 1980. Not only is this leading to the extinction of millions of species residing within the forest, it also adds to the CO_2 problem in the atmosphere.

Carbon dioxide, in excess, harms the only thing keeping solar rays from overheating the Earth: the ozone layer. As humans add more CO_2, the ozone layer is destroyed further. If humans continue like this, in a couple hundred years humans may not have a world to look forward to.

However, we could reverse our actions by planting more trees, using fewer fossil fuels, and enacting laws to enforce environmental protection. Also, we need to bring more media attention to the causes and problems of global warming to inform the rest of the world. Change won't happen immediately, but over time we can reverse the effects to an extent where it's not dangerous. The fate of the human race depends on it.

Stephanie Lam *was born in Hong Kong and moved to San Francisco when she was seven. She plays basketball on the high school team, practicing for two hours each day! She likes to peruse YouTube videos, especially those about Japan because she likes manga and anime. When she was in kindergarten, she broke her arm going down a slide. She has no idea how that happened. She hopes to study computer science and engineering in college.*

THE MYSTERIOUS SPECIES

Coral reefs are very colorful and diverse ecosystems. They have many colors, like the artist Leonid Afremov's paintings. The variety of species relying on the coral reefs is like the diversity of race in San Francisco. I chose to write about the destruction of coral reefs and the ecosystem in the Coral Triangle because it can actually affect the species and people that live in that ecosystem, and it can also affect people and the economy worldwide. Biodiversity is important to an ecosystem because the more species there are, the more stable the ecosystem is. The other reason I chose to write about this is that not many people know about what is happening to the coral reefs. People don't think and don't know how their actions are destroying the coral reefs. Coral reefs help protect coastal communities from storm surges and erosion from waves. Both are likely to increase in the face of sea-level rise. Many medicines have been obtained from coral reef organisms, according to *reefresilience. org*. Humans are destroying coral reefs and their ecosystem, so humans should be more careful in their actions.

How are humans destroying coral reefs? Humans are destroying coral reefs in both direct and indirect ways. Direct destruction is when human activity affects the reefs directly and an example is overfishing. Overfishing is an example of direct destruction because people are taking out a lot of fish from the reefs. Meanwhile, global warming is an example of indirect destruction. Humans are adding carbon dioxide into the air, which causes the water temperature to rise, which later affects the reefs.

Coral reefs are very important. Coral reefs are built by coral polyps and require zooxanthellae algae, which is found inside hard corals. Some abiotic factors that the coral reefs need to survive are sunlight, clear water, clean water, saltwater, and warm water temperature. For example, if the water temperature gets too high or low, then the coral reefs would die. Coral reefs can be found in more than one hundred countries around the world and cover about 110,000 square miles worldwide, according to the Coral Reef Alliance. Small coral fish live in coral reefs and use them for protection. People use coral reefs to attract tourists and help bring money to the economy. Even though coral reefs can be found in a lot of places, the number of coral reefs is dropping due to the destruction and damage being done by humans. More than eighty-five percent of the reefs in the Coral Triangle are directly threatened by local human activities like fishing, according to *wwf.panda.org*.

The Coral Triangle is a marine region that spans across Indonesia, Malaysia, Papua New Guinea, the Philippines, the Solomon Islands, and Timor-Leste. The Coral Triangle contains two biogeographic regions, and they are the Indonesian-Philippines region and the Far Southwestern Pacific region. The Coral Triangle sustains 120 million people. Some of the species that live in the Coral Triangle are whales, dolphins, tunas, sharks, sea star, and many more, according to *wwf.panda.org*.

Humans are affecting the coral reefs and the ecosystem in the Coral Triangle in many different ways. Global warming, climate change, and overfishing are the main causes for the destruction of the coral reef. Global warming and climate change are causing water temperatures to rise and coral reefs can only live in warm water temperatures. Climate change also impacts this ecosystem. It warms and acidifies the water and causes seas to rise.

El Niño, an event that affects the weather patterns in the Pacific Ocean, also impacts coral reefs. El Niño causes water temperatures to rise for a period of time and that leads to coral bleaching. Coral bleaching is when the coral reefs get exposed to warmer-than-normal water temperatures and excessive sunlight for about four to six weeks. Coral bleaching can kill the coral reefs by making the plant-like cells in them toxic, which makes the coral reefs starve to death.

Some of the consequences from global warming and climate change are more intense weather, acidic seas, rise in water temperature, etc. More intense weather can destroy the coral reefs and their ecosystem. Acidic seas can cause the coral reefs to collapse and die. If the water temperature gets too warm, the coral reefs won't be able to survive.

Overfishing is affecting the coral reefs because some fishermen use destructive fishing methods to make fishing large amounts of fish easier. Some destructive fishing methods are using explosives, bottom trawling (towing or dragging a fishing net along the sea floor), Muro-Ami (beating the reefs with rocks to scare the fish inside the reefs), and many more. Some of the species being overfished are tuna, shark, billfish, and many more. Fishermen often use illegal methods to fish large amounts more easily.

Competition between fishermen leads to overfishing and eventually to the extinction of fish. Some consequences of overfishing are killing the coral reefs, species being killed accidently, etc. One effect of overfishing is tragedy of the commons, because when fishermen overfish, the fish (common resources) can't reproduce fast enough and then the fish end up extinct, which is the tragedy. All the destructive fishing methods kill the coral reefs and other species. Overfishing can cause imbalance to the ecosystem because the big fish will be extinct, leaving the small fish behind. About $23.5 billion of global finances is lost due to illegal fishing, according to *worldwildlife.org*.

There are many organizations that are trying to save the coral reefs and raise awareness about them. Some of those organizations are Coral Triangle Initiative, Coral Reef Alliance, World Wide Fund for Nature, and many more. Some of the possible responses can be making laws or regulations on overfishing and teaching fishermen about the causes and effects of overfishing.

The most modern and technologically advanced countries in the world should make laws and regulations to slow down global warming and stop adding too much carbon into the atmosphere. Unfortunately, there are many factors that stop the most modern and technologically advanced countries from making laws and regulations regarding global warming. Some of the factors are private corporations, factory owners, and business owners. These people are saying no to laws and regulations to slow down global warming. Some people are saying global warming is not happening or saying it is not real. Some things people can do to slow down global warming are carpooling, riding public transportation, etc. Adding carbon into the atmosphere causes water temperatures to rise, which will destroy the coral reefs. We should do those things to slow down the process of adding carbon into the atmosphere. It is important to save coral reefs because saving them can help save the economy, the species, and the people.

Denis Aguilar *is from Santa Tecla, El Salvador, and moved to San Francisco in 2013. He likes to play soccer and is the team captain. His favorite food is pupusas. He wants to be a doctor, and when he goes to college, he will be the first in his family. But first he wants to learn more English.*

HOW EVERYBODY CAN HELP EACH OTHER

According to the Public Policy Institute of California, "California is home to more than 10 million immigrants." One way immigration contributes to California's strong economy is in jobs. California's economy is an ecosystem because it has biotic factors like immigrants and U.S.-born citizens, and it also has abiotic factors like jobs. Equilibrium is maintained because the more people who are working, the stronger the economy is. If either immigrants or U.S.-born citizens are unemployed, the economy is disrupted. Immigrants benefit everyone in the ecosystem, both people in California and those in the immigrant's home countries.

I learned about immigration at school and from my family and some of my friends. This topic is very important because a lot of immigrants are sent back to their countries through the Immigration and Customs Enforcement, and some of them feel scared because of the violence in Central America or in other countries. The equilibrium in my social community was disrupted when my family told me that I was coming to the United States. I felt weird because I didn't want to leave my family, friends, and my country, El Salvador. I felt weird because I knew that everything was going to be different. For example, friends, language, and my mom, because I didn't know her that much, or even how she looked. When she came to the U.S., I was three years old and I didn't know her.

Immigrants are important to California's ecosystem and immigration benefits. California has a large percentage of immigrants: twenty-seven percent of the population in California was foreign-born in 2013, about twice the U.S. percentage (Public Policy Institute of California). This twenty-seven percent of foreign-born people are like me because all of them come to California to have a better life and better education. One reason that the percentage of immigrants not born in the United States is high in California might be because California has good benefits for immigrants, like jobs. For example, my mom came here to California because of these benefits.

Immigrants to California come here to have a better life. Fifty-three percent of immigrants in California come from Latin America. Many come because of the violence in their countries. They could also come to help their families because some countries are poor and the way that they help give opportunities to their families is by working hard. One example is me. I came from Central America to California because in California I can have more opportunities to go to college and get a really good job to get money and help my family in El Salvador.

Immigrants are more likely than U.S.-born residents to be employed but they make less money. Sixty-six percent of immigrants in California are employed compared to sixty-two percent of U.S.-born Californians. Some people might think that the employed immigrants and U.S.-born people are like parasitism because the immigrants are benefiting by working more than U.S.-born people. But, other people like me might think that it is like mutualism because both are winning by working and getting money and contributing to the economy of California. When immigrants come to the United States to work they make more biodiversity. They come because they want a better life and they get used to the new life and new things. They introduce new items, such as food into the ecosystem, making it stronger.

I hope in the future that I can be part of the percentage of immigrants working in California, and that I can make the economy stronger and help my family like most immigrants help their families.

Daniela Sandoval *is a born and bred San Francisco girl. She has a real love for and commitment to math. In the summer of 2015, she participated in the SMASH (Summer Math and Science Honors) Academy at Stanford; she'll be participating again in the summer of 2016. She learned to swim in the first grade. She is the only girl in a family with four boys, and plans to go into a career in finance.*

LET'S DO EVERYTHING WE CAN

How would you feel if a doctor told you that you had limited time to live? Wouldn't you do everything in your power to extend that time, even if it meant a little sacrifice? California is coming to the point where we are running out of resources, more specifically water, which is essential to life and that is limiting our chance of survival. We humans have the power to extend those resources and help increase our chance of survival. In order to conserve our water resources, we should use less water for agriculture by growing only what is necessary. This will not only reduce water use, it will also diminish waste and could help improve our drought issue.

California is an ecosystem that includes both biotic (living) and abiotic (nonliving) things. Humans, animals, and plants are some biotic factors that are both affecting and being affected by the drought. The major abiotic factor being affected in this case is water. The maximum amount of food our ecosystem can support depends on our water supply because in order to grow food, water is essential. "California is entering its fourth year of a record-breaking drought," says *CBS News*. The historic average rainfall for a five-year period in San Francisco is 118.25 inches, and the actual rainfall from the past five years has significantly decreased to just 72.37 inches. Another city that is affected by lack of rainfall is Los Angeles; its average rainfall has also drastically gone down, from 74.65 inches to 35.65

inches. Without water, we can't grow food. According to *The Washington Post*, eighty percent of California's water is used for agriculture.

California is wasting water without even knowing it. Eighty percent of fresh water is used for agriculture and agricultural regions, which include farms that grow our food. After growing the food, sometimes the food gets very ripe and the sellers do not want to put it up for sale and instead they throw it away. The food that gets thrown away is mostly still edible in my opinion, but the stores don't sell it so their stores won't look bad. They only sell the perfect-looking produce. According to the Natural Resources Defense Council, fifty-two percent of the food grown in the United States goes to waste, so in California that would mean that fifty-two percent of the eighty percent of water that is used for agriculture. This means that a total of 43.2 percent of water in California could be going to waste through agriculture. This is an example of the tragedy of the commons. California's agriculture industry is taking more water, a natural resource, because it is in its own interest, but at the cost of a drought in which we are all affected negatively. Farmers are growing as much food as possible so they can make profit off it when they sell it. They are being selfish because they know we are in a drought, but they act this way for their own interest in making profit. If we grow less food, then less food would go to waste, meaning less water would go to waste. With just enough food and water resources, we could increase our carrying capacity because there would be enough resources for everyone for a longer period of time. Instead of throwing away good food and letting it go to waste, we can make sure that we use up only what we need.

A significant portion of California's crops get exported to other states and countries, and farmers are getting money in return. What if we exported fewer crops and only produced what is needed here in California? It would affect the farmers negatively because they would be making less money, but in return we would be saving a portion of water that would expand our water and food resources. This would be an example of a mutualistic relationship because by taking action and growing less food, farmers would be giving back to the community by saving more water, and they would be saving water for themselves as well. With more water, we all have higher chances of surviving, eating well, and reproducing.

But predation is another ecosystem relationship that could be taking place in this situation. People with more money would have more control in this case, and with more power they can take advantage of others with some decisions they make. For example, if a farmer is making lots of money by selling his crops, and we are asking for him to grow less food, he might not want to because of the ton of money he is making. In this situation, the people who don't have control of the agricultural water supply are affected negatively because they are being deprived of water. Farmers sell what they grow and get money in return, but we really should be worried about the water for us all. Our lives depend on it.

California is in a drought and water is very important. We should value what we have and use it wisely. Farmers should use less water to help save California and all organisms who live here from the severe drought we are in. We, as inhabitants of California, should watch out for ways we may be wasting water, and food as well. Let's make a change and help save California.

Samantha Gomez *is a San Francisco native and a sophomore at John O'Connell High School. When she's not in school, she's busy volunteering in her neighborhood, playing football with friends, and making really good pizza from scratch. She wants to go to college and be a doctor. This is her second time being a published author.*

HOT SKIES

Humans are a big cause of global warming in San Francisco and the Bay Area, but we can also be the solution. Burning fossil fuels causes global warming, by creating greenhouse gases which trap heat in the atmosphere. When the earth gets warmer, the icebergs start to melt, which causes sea levels to rise. This affects us because the sea-level rise will one day cause overflow, and it will start flooding parts of San Francisco. This is an example of the tragedy of the commons. People think about their own benefit when using their cars because they might be too tired to go to a bus stop, but in the end, it's going to be bad because they are burning fossil fuels and it's hurting the environment. Tragedy of the commons means a situation where individuals act in their own best interest but at the cost of the group's success. Everyone involved is eventually hurt.

The San Francisco Bay is an ecosystem because it is made of biotic and abiotic factors. Biotic factors are living things, for example: humans, fish and sea animals. Abiotic factors are nonliving things like water. These factors interact with each other in an ecosystem. Fish (biotic) need water (abiotic) to survive. The warming climate changes the ocean to such a degree that the survival of seals and their young has increasingly become a concern for marine biologists. The loss of sea ice in Antarctica has caused a decrease in the amount of algae, plankton, and krill, the foundation of the ocean's food chain.

The rising sea level is really bad for us. The rising sea level causes waters to overflow, which can be a dangerous thing. If the water were to rise two meters in San Francisco, parts of North Beach and 3rd Street all the way to parts of Cesar Chavez Street would be underwater. If water were to overflow two meters, half of Oakland would be gone. This would cause people to be displaced and the homeless to suffer the most. This all happens because of global warming, which humans cause. We don't think about this, but it's real life and it's scary that this can happen if we proceed to do the bad things we are doing to cause climate change and global warming.

The best way to solve the problem is to use more "green" cars. A hybrid car gets better mileage than a traditional gasoline-powered vehicle by five miles per gallon on average. And, because hybrid cars run in part on different fuel sources, they do not need to be filled up with gasoline constantly, cutting overall gasoline consumption and cost. When gasoline is burned to produce energy for vehicles, carbon monoxide is created and released into the environment. However, with hybrid cars, much less of this bad substance is released into the atmosphere. This is reducing emissions and air pollution.

I believe that we all should start caring a little more and ask ourselves, *Should I use the car today?* and try taking the bus. We need to think twice about everything we do and ask, *Will this hurt the San Francisco Bay somehow?*

Ricardo Gallardo *enjoys playing video games and basketball (he's a former LeBron James fan), and also enjoys and excels in history, especially ethnic studies. He's interested in how past oppressions have affected future generations and how inequalities still remain, most notably in his own San Francisco Mission District neighborhood. He dreams of becoming a chef and working with fast American cars in drag racing.*

THE ROTTENING OF THE MISSION

According to Carol Pogash of *The New York Times*, "Luxury condominiums, organic ice cream stores, cafes that serve soy lattes and chocolate shops that offer samples from Ecuador and Madagascar are rapidly replacing 99-cent stores, bodegas and rent-controlled apartments in the Mission District, [San Franciso's] working-class Latino neighborhood."

Basically, in scientific terms, the "techies" who are moving in to San Francisco to take jobs in tech, are the stronger organism. They have more money, and money talks. They also have more education than some people in San Francisco's Mission District. The weaker organism is us, the people of the Mission. We don't have money like they do, but in a way we are stronger because we have more people and stronger cultural ties with this neighborhood.

When I was younger, the Mission used to be very different from the way it is now. The Mission wasn't always filled with techies. Before techies decided to come into the Mission, it was a scary place. I remember hearing about a shooting when I was younger. I was probably five or six when this happened, but I heard my mom talking about how they shot a guy dead in the head on the corner of 20th and Mission, right across from Pete's BBQ. A few days after I heard about that, my mom and I were on that street and

I looked down at the spot and I saw dried blood and told my mom, "Look, that's where the guy got shot."

Now that the police have calmed all the gangs, techies have decided to come. To me, the Mission used to be somewhere I would only go with my parents since it was dangerous and I was young, but even in that state, the Mission was a beautiful place. It showed a lot of culture. Back then you would actually see people hanging around, like a group of adult friends sitting outside their house, just hanging out playing cards or dominos with a radio blasting music. Nowadays that's not something you come across. You see people hanging around the Mission, but that's because they don't have jobs. They have nowhere to go so they just have to be around the Mission.

In the state that the Mission is in right now, it's kind of like parasitism, because the only organisms that are winning are the techies. What is that costing us? It's causing people like us to lose our money, our houses, and our jobs. Is there any hope for a mutualistic Mission?

"The affordability crisis is so extreme that many of those who rode into the Mission District on the first wave of gentrification, during the dotcom boom in the 90s are now crying foul. Even they can't afford the two-bedroom apartment on Valencia Street renting for $11,500 a month. They find themselves priced out of their lofts and community networks, by a whole new wave of highly paid tech workers who ride in on the Google bus every evening, driving rents and home prices to dizzying new heights," writes Maria Poblet at *Causa Justa*.

"The average asking price in San Francisco's central business District (CBD) is now about $66 per square foot," according to property management company Jones Lang LaSalle. "Five years ago that same square foot cost about $36. Office rents here have soared 83 percent in five years, nearly four times faster than downtown New York," says Brett Murphy from CNET. The prices for office space in San Francisco are rising a lot by the square foot, which causes companies that were already here to move somewhere else, like Oakland. Also, nonprofits are having to move along with them. A lot of techies are all moving in from near Palo Alto because they didn't like living in that area. The companies decided that for the people who do live in the city, a big charter bus would pick them up and take them to their job in Palo Alto and bring them back at the end of the day.

One reason why so many people decided to come to San Francisco is because this city has always had ethnic diversity. That's something this city has always been known for. Studies say that the white population is on a steady increase. Studies suggest that by 2040 there will be more white people than people of color. The way that the Mission can be equal is obviously mutualism, but that's something that will not be happening anytime soon or easily. Since the techies are moving into the Mission it's not easy to reject their money. A house that was once $30,000 is being bought for $1 million or more; landlords and homeowners can't resist that type of money. They're getting a huge profit from selling their house. When I walk around the city, I see a whole bunch of abandoned buildings and wonder, *Why don't they use these empty buildings instead of taking peoples' homes?* I think that if they can turn an old neighborhood that used to be warehouses into luxury condos, they shouldn't have a problem finding random abandoned buildings and using them for a better purpose than a luxury condo.

Janet Chen *hails from the mist-soaked avenues of the Richmond District at the very edge of San Francisco. Jane's life currently centers around pursuing her interest in math, passing the time at John O'Connell High School, and returning home each day to absorb herself in* Smallville *and* Jane the Virgin.

SEA LEVEL'S RISE MAY BE ISLAND'S DEMISE

One day last year, a family on the Marshall Islands in Micronesia woke up to water flooding their home. The water was coming from the ocean. This would not have happened twenty years ago.

On a daily basis, an average person will most likely not think about disappearing islands. They probably won't even think about global warming. I know I don't. Most people just care more about other things. For example, approximately 100 million people watched Super Bowl 50 and could not wait for the halftime show. The chances of them thinking about global warming during the game were unlikely. This just shows that many people disregard global warming and its effects on islands that are out of sight. I believe that if people cared about global warming as much as they care about the Super Bowl, climate change wouldn't be a huge problem as it is now.

Scientists have confirmed that the sea level has risen due to global warming. There is more carbon dioxide in the air, trapping heat on Earth. This causes the ice caps and glaciers to melt and the sea levels to rise. As a result, many islands are going underwater. Specifically, global warming is harming the Marshall Islands, and unless we make laws to reduce the carbon footprint of companies around the world, the Marshall Islands will soon disappear.

Not only should we make laws, but we must also take individual steps that can slow down global warming.

The Marshall Islands are located between Hawaii and Australia. The average height of the Marshall Islands is just seven feet above sea level and together they have a total area of about seventy square miles. There are many living things on the islands, such as humans and animals. The islands have forests, which provide food, materials for houses, medicine, and many other items. There are also coral reefs. People on the Marshall Islands have begun to notice coral bleaching, which happens when warmer water temperatures cause coral to turn white. Coral can handle bleaching occasionally, but too much can be dangerous. Coral is crucial to the islands because it forms an important barrier from the ocean waves. Developed nations are causing this because the carbon footprint they leave behind is affecting the ocean water temperature.

Majuro is the capital city of the Marshall Islands, and Kwajalein is another island located northwest of Majuro. The average annual number of warm days in Majuro has risen by twenty-five days since 1960, and the average annual number of cool nights has decreased by forty-five. Meanwhile in Kwajalein, the average annual number of warm days has increased to 110 days since 1960 and the average annual number of cool nights has decreased by eighty days. This is important because when ocean water warms up, it expands, causing the sea level across the world to rise. According to the Australian Bureau of Meteorology, "The sea level has risen near the Marshall Islands by about 0.3 inches per year since 1993. This is larger than the global average of 0.11–0.14 inches per year."

This is because of human activities. According to the Union of Concerned Scientists, Chevron, ExxonMobil, Saudi Aramco, British Petroleum, Gazprom, Shell and the National Iranian Oil Company produce almost one-fifth of all the industrial carbon in human history. These companies and many more are satisfying their customers and just looking at their own best interests to make profit. Furthermore, they are hurting the environment by polluting the air. In the long run, they are also hurting themselves. This phenomenon is known as the tragedy of the commons.

Power plants also release tons of carbon dioxide. According to the U.S. Energy Information Administration, the U.S. released a total of "2,043 million metric tons, or about 38% of the total U.S. energy-related CO_2 emissions in 2014."

In December 2015, 195 countries met in a climate conference in Paris and reached an agreement about climate change. According to the European Commission, "The agreement sets out a global action plan to put the world on track to avoid dangerous climate change by limiting global warming to well below 2°C." The U.S. is one of the countries that participated in the agreement. However, the Supreme Court has halted President Obama's decision to follow the agreement. This is not right because many other countries have committed to cut down on their greenhouse gas emissions. It would not be fair if the U.S. goes against their word.

In addition to the government doing its part to reduce global warming, individuals need to help as well. A few things they can do are to purchase a fuel-efficient low greenhouse gas vehicle, take public transportation, bike, recycle, and even replace a regular incandescent light bulb with a compact fluorescent light bulb.

The Marshall Islands are disappearing because developed nations have cars, industries, and power plants that release tons of carbon dioxide into the air. It is of great importance that we work together as a nation and as individuals to reduce our carbon footprint to stop the islands from disappearing for good.

Katherine Torres *was born in Usulután, El Salvador, in 2002. In 2002, she moved to San Francisco to be with her mother who had immigrated to the United States three years earlier. Katherine wants to go to university to become a pediatrician.*

NO RAIN, NO JOBS: THE CALIFORNIA DROUGHT'S EFFECT ON IMMIGRANTS

The California drought affects not just California's citizens, but also people across the United States. One of the major reasons is because California grows more than one-third of the country's vegetables and two-thirds of the nation's fruits and nuts, according to the California Department of Food and Agriculture. This is especially hard on Latino immigrants, because many of them work in the fields at minimum wage jobs, but we can change that by encouraging all California residents to conserve water.

Latino immigrants are suffering because the drought is causing water prices to increase. An example of how the drought is affecting immigrants is my mother, who emigrated from El Salvador about ten years ago. She told me that when she opened her water bill and saw the big rate increase, she got as mad as the Grinch. She gets frustrated because the increasing rates are affecting her finances. She doesn't just need money to live in the U.S. and pay for all our expenses, she also needs money to send to our family in El Salvador to cover their expenses, too.

Another financial effect of the drought is that people in agriculture are losing their jobs because the drought is killing the plants and vegetables. Currently eighty percent of the water in California goes to water agricultural fields, according to *The Washington Post*, and many Latino immigrants in California work in the agricultural industry. Because the drought means the farmers cannot grow as much, the immigrants are losing their jobs and many are unable to feed their own families because there is not enough money. This is an example of California reaching its carrying capacity, which happens when there are too many workers, but not enough jobs to go around. Just like any ecosystem that supports a certain amount of species, the system will be unable to continue and will break down. The agricultural workers' ecosystem will suffer when there are not enough jobs.

If all people in California conserve water, it will help to decrease the effects of the drought. Every day, most Californians, including immigrants, also find new ways to conserve water. Like you, like me, people in California are taking shorter showers, not watering their lawns, growing drought-tolerant bushes, and collecting the water that they don't need from the shower and laundry to use elsewhere in the house. Unfortunately, not all Californians are willing to make the effort and the sacrifice of conserving water. These people continue to wash their cars and water their lawns. Their refusal to conserve is called tragedy of the commons, a situation where individuals act in their own best interests, but at the cost of the group's success. For instance, everyone knows that they need to conserve water, but some people are bad citizens, or water hogs.

As an immigrant, I'm aware the drought has worse financial effects on my fellow Latino immigrants. This is because immigrants generally have lower paying jobs and they can be very susceptible to the weather, especially the drought. This is a very serious problem that is affecting all Californians. Whether we have a good year or not, I encourage all of you to conserve water instead of wasting it. Let's make a difference in this current situation, because if we don't start conserving water, the drought will always be with us.

Julien Nelson *is a sophomore at John O'Connell High School. He moved to San Francisco at the beginning of the school year from Jersey City, New Jersey, with his mom, two giant cats, and his Keeshond puppy. Julien's favorite subject is math, and he hopes to work for a tech or video game company after studying computer science at either UCLA or UC Berkeley.*

THE INDUSTRIAL PROBLEM

The Industrial Revolution started an evolution of technology, along with an evolution of problems. The Industrial Revolution affected the ecosystem of the world's largest city at the time, London. An ecosystem is a community of biotic factors, being people living in London, and abiotic factors, being machinery interacting with each other. Between 1760 and 1840, people had new ideas and inventions. They came true due to resources like coal and iron, which started the boom of the Industrial Revolution. Even though the Industrial Revolution started the new generation we have now, it also started environmental pollution, overpopulation, and other problems that we currently have. In eighteenth century London, machines in factories created pollution in the sky and open sewage contaminated water. Contaminated water caused disease, and overpopulation accelerated all of these issues. The Industrial Revolution had inspired pollution and mass production in eighteenth century London, which had a negative impact on the environmental ecosystem and continues in our world today.

The city of London became busy and populated due to more work and better pay during the Industrial Revolution. People migrated from Ireland to feed their families due to a potato famine. Farmers left their farms to work in the cities because they were replaced by machines in the fields. The Industrial Revolution was a time of inventions, and these inventions required energy from burning coal, which caused pollution. Burning coal causes air pollution by releasing sulfur from coal that combines with oxygen. It creates sulfur dioxide, which can cause several different negative effects. When sulfur dioxide (SO_2) is released into the atmosphere, it can combine with water droplets that make up clouds. They can create acidic rain that can harm trees and animals. Air pollution also causes particulate matter, which could be dangerous depending on how long you are exposed. Effects range from irritated eyes, allergies, and headaches, to cancer, heart disease, and damage to the brain, nerves, and so on.

Air pollution was not the only problem from the Industrial Revolution. There was also water that caused disease. In London there was open sewers where people threw their literal crap out the window into the sewage. Little did the people know, sewage came in contact with the drinking water, which spread disease. Most of London used the river as their drinking water, which was contaminated and made the diseases spread quickly. Diseases such as cholera, tuberculosis, and typhoid are all major diseases transmitted through water. At the time, there was confusion about what caused disease, according to *historylearningsite.co.uk*. "Even a great reformer like Edwin Chadwick was convinced that disease was carried in the atmosphere which had been poisoned by foul smells," even though the disease was spread by water, according to the website.

Within 250 years, things like water pollution still go on in a country with advanced technology and money like the United States. Flint, Michigan, had switched the supply of water to save money, which caused health problems because the water got contaminated with lead. This is an example of a tragedy of the commons in which the government lied about the water being safe just to benefit themselves by saving money. The Industrial Revolution did the same by valuing profit over the safety of the ecosystem. Mass production and population have evolved even more since eighteenth century London. The tragedy of the commons has grown to a global scale. Since the eighteenth century, factories have wanted to make more goods for profit, which requires the use of even more resources, which contributes to more air and water pollution. We as people, in order to change, should be more aware of the consequences that affect the environment and our ecosystem for a better future.

Our ecosystems all around the world are having problems with pollution, tragedy of the commons, and disease. In some places like India, there are still open sewers and disease that affect peoples' lives everyday. Major companies use child slaves without consumers knowing in order for companies to have a higher profit. In order to fix this, we need to be more cautious of our actions, understand the consequences that we have on the environment around us, and not ignore it. For example, we should be less greedy with our resources and technology so we can share it with the world in order to make it a better place.

Erick Arevalo *was born and raised in San Francisco, California. He enjoys playing* Halo V. *In addition to video games, Erick enjoys watching his favorite team, Barcelona Football Club, and cheering for his favorite players, Messi and Neymar.*

ENVIRONMENTAL DAMAGE IN THE SAN FRANCISCO BAY

I've been living in San Francisco since I was born. I always tell my brother not to litter, because he litters all the time. I do not like to see my brother or other people litter because it's bad for the ecosystem. An ecosystem is an environment where biotic factors and abiotic factors exist. The San Francisco Bay is an ecosystem. Some examples of biotic factors of the Bay Area are the fish and the humans that live by the Bay. An abiotic factor in the Bay is the litter. The interaction between the biotic factors and the abiotic factors can cause danger to the Bay by making it dirty. This is bad for the ecosystem because the fish could get infected. If the fish get infected, humans get infected.

According to *The San Jose Mercury News*, each day people from the Bay Area litter 3.9 million pieces of plastic without even knowing it. They litter tiny pieces of synthetic fabric from clothes and they get broken down. The plastic does not only contaminate the Bay, but the fish get contaminated. People catch the fish and eat them, which is a health risk. In comparison, other cities north of Oakland and San Francisco drop about 310,000 pieces of litter. That's not even a quarter of what we, the people that live by the Bay, drop daily. Plastic, or litter, also comes from storm drains, creeks, rivers, and illegal dumping. The fish eat the plastic and the humans eat the fish, so then humans can get an infection or a disease from the fish they eat. According to the Natural Resources Defense Council, chemicals called *phthalates* disrupt testosterone and other hormones, and "those phthalates can leach from water bottles into the Bay." Another effect of plastic in water is that it can cause cancer. Littering is bad because it has many health risks, and it's bad for the ecosystem because it affects fish and humans.

The ongoing problem is who to blame between the city and the people; it will end with both of them deserving blame. People drop the litter, and the city does not protect the Bay from the trash that gets into storm drains. The city should make nets that cover storm drains, so that trash can't get in them and wind up in the Bay. Humans can change their behavior by not throwing any trash into the sewers that connect to the Bay.

If you see any trash on the floor, pick it up and throw it away. By throwing trash in trashcans, we stop fish and humans from getting sick. Eating fish that have eaten plastic or that live in polluted water can lead to cancer and other dangers, like abnormal hormone levels. By throwing away trash in trashcans, there's less of a chance of fish getting affected. City regulations on plastic water bottles would also help the Bay because there would be fewer water bottles. Fewer water bottles would mean that fewer fish would be affected by plastic, and humans would be less affected when they eat fish. It's important for humans to change their behavior, because they can affect more than just fish; they can affect a whole ecosystem.

I think that the best way to solve the problem is to change to reusable food and drink containers. This way the fish would not get a disease from the plastic and the city would not be affected. According to *cleanwater.org*, we can stop littering up to forty percent of the trash by eliminating plastic food and drink containers. If we had reusable food containers, we would be reusing the containers and wouldn't throw a lot of trash in the sewer.

In conclusion, the San Francisco Bay is a target for litter. The litter affects it by contaminating the fish in the Bay. The contamination can lead to human disease by people eating fish that ate the litter in the Bay. These health risks can lead to cancer and other dangers like abnormal hormone levels. Because of these effects, we, the people that live around the Bay, should stop throwing litter into the Bay to stop dangers and health risks. The Bay is my home, and I don't want it to end up like a piece of junk, with the whole ocean getting polluted and turning dirty. I don't want that because I grew up here and I plan on living here for the rest of my life.

Daniel Ceron *is fourteen and was born in San Francisco. He loves soccer and is a forward-striker on his club soccer team. In five years, Daniel aspires to be playing in the El Salvadorian soccer league, and then maybe move on to Major League Soccer. Daniel likes watching movies, and his favorites are* The Avengers, The Incredibles, *and* The Maze Runner.

THE TRAGEDY OF 9/11

It was only 8:46 a.m., and the tragedy of the commons had happened! Taking American Airlines flights 11, 175, 77, and 93, terrorists attacked the United States. Since they didn't have enough power to attack the U.S. military, the terrorists attacked something else that symbolized the United States. They attacked two places: the Pentagon and the World Trade Center, also known as the Twin Towers.

These nineteen terrorists attacked the U.S. to frighten everybody. Some 2,977 people died in the attack known as 9/11. This action caused huge amounts of human fatalities in the resulting wars between the United States military and the Middle East. In this case, the success of the nineteen terrorists from al-Qaeda caused 2,977 people to die, including the terrorists. In biological terms, an event of this nature—where individuals are acting in their own best interest but at the cost of the group's success—is called the tragedy of the commons.

Osama bin Laden planned this attack against the United States. Bin Laden and the nineteen terrorists are all connected with the group called al-Qaeda, located in Afghanistan. In this quote from a speech by bin Laden, there is evidence of what bin Laden thought about the United States:

No, we fight because we are free men who don't sleep under oppression. We want to restore freedom to our nation, just as you lay waste to our nation. So shall we lay waste to yours. No one except a dumb thief plays with the security of others and then makes himself believe he will be secure. Whereas thinking people, when disaster strikes, make it their priority to look for its causes, in order to prevent it happening again. But I am amazed at you. Even though we are in the fourth year after the events of September 11, Bush is still engaged in distortion, deception and hiding from you the real causes. And thus, the reasons are still there for a repeat of what occurred.

This quote shows that the reason why al-Qaeda attacked the United States was because bin Laden wanted revenge, because he believed that the U.S. was attacking his people first.

The people who are in this ecosystem are the United States and al-Qaeda. They are both from different regions, and they both have different ideologies. Their differences are abiotic factors. The definition of abiotic is something that does not have life. For example, in the 9/11 attack, the planes were abiotic. Other abiotic factors were: religion, buildings, weapons, and government structures. The definition of biotic is the opposite of abiotic: it is something that has life. For example, the terrorists, all the people who died, all the leaders of both involved groups (bin Laden and George W. Bush) are all biotic factors. These abiotic and biotic factors interacted in the attack on the United States, and this resulted in the tragedy of the commons. The nineteen terrorists were the group that succeeded, but they all eventually died.

Some other examples of the tragedy of the commons that happened was the United States government using so much money to rebuild, but failing to fix the economy. At the end of the attack, a whole building fell down, so they had to use money to fix the damage. Both the Twin Towers and the Pentagon cost money to repair and replace. After the attack, the U.S. also spent money on the National September 11 Memorial and Museum. The money that was lost to 9/11 is an example of a tragedy of the commons because the government succeeded at rebuilding, but it took so long and cost so much that it lost a lot of money and ultimately, even though the government succeeded at its goal, everybody lost.

The definition of altruism is sort of like the opposite of tragedy of the commons. It's when individuals help the greater good, but they lose something. For example, in this case the New York firefighters who sacrificed themselves to save others during the 9/11 attack used altruism to save others. Other evidence of altruism was when the people who were in the plane used altruism to save others. Another example of altruism is the terrorists themselves. They sacrificed themselves to help al-Quaeda achieve its goals. "Al-Qaeda hoped that by attacking these symbols of American power, they would promote widespread fear throughout the country and severely weaken the United States' standing in the world community, ultimately supporting their political and religious goals in the Middle East and Muslim World," according to "FAQ About 9/11." In this quote, you can see the tragedy of the commons and altruism happening in one ecosystem.

One thing you guys might have learned reading this essay is that ecosystems are not only for animals. An ecosystem can also be the world, or it could be about regions. It could be about a lot of stuff. While writing this essay, I learned that terrorists could be anywhere, but you can't judge them or judge Muslims. You can't judge a book by its cover.

Why doesn't al-Qaeda want to attack the U.S. again? Is it because bin Laden is dead and they don't have a leader? Writing this essay got me curious. I wondered whether al-Qaeda had planned this for decades or years. And what if they are planning another attack? If they are planning another attack, I think that the U.S. should already be prepared, but nobody knows whether we are except the people in the government.

For us to be prepared for something to happen, we should have extra supplies at home, for example, a first aid kit, food, water, and a blanket. We should start planning on this day, because we never know when this could happen. It could probably happen while you are reading this.

Zykeill Council *is sixteen years old and has lived in San Francisco his whole life. He likes to be outside enjoying nature, playing basketball, and driving. He lives with his auntie and two little cousins on Treasure Island. After high school, he wants to go to college out of state and study biology.*

HOW TO SAVE THE BAY

My dad and I were outside in front of our house near the porch, washing my father's car, a silver 2014 Ford. When we were washing the car, my father took out a piece of candy called a Chick-O-Stick. He tossed the wrapper onto the ground. It was then washed away as a boat would be by a big wave. That's when I thought to myself, *This is probably how the littering and pollution cycle keeps going.* How is littering polluting our bay? The trash thrown on the ground gets flushed to the San Francisco Bay, and it can be eaten by species and hurt their bodies.

The ecosystem in the Bay is very diverse and full of energy from animals like otters, seals, and plankton. The ecosystem is made of a bunch of abiotic and biotic factors that help the ecosystem. Some are also detrimental to the ecosystem, such as the trash of the people in the city who are littering. Some other abiotic factors in the Bay are oxygen, rocks, and seashells. Many things are littering the Bay, but the most abundant item littering the Bay are cigarette butts. The second most littered thing in the Bay is polystyrene, which does not biodegrade. Researchers discovered that fish in the San Francisco Bay had an average of six pieces of plastic inside of them when they accidentally caught nine fish as they gathered water samples, according to *contracostatimes.com*.

Littering is an example of tragedy of the commons which has an effect on the ecosystem of the Bay. Some people litter because they are tired of holding trash, and in the short run, they get rid of the trash and they are acting in their own best interest. In the long run, they hurt fish and other biotic factors in the ecosystem. They also are hurting themselves because they are limiting a food source for many other beings.

There are many ways we can prevent littering and the pollution of our environment. The city can help by placing dozens more public trashcans around the city. This way people have more options as to where they can throw their trash. The community can help stop littering by not doing it themselves and by encouraging their peers or youth not to litter. Law enforcement and the government can stop people littering by making a law so you can get a $500 ticket if you litter.

One of the most productive ways of stopping littering, in my opinion, is the street cleaning process. The street cleaner actually cleans up excess oils, road sand, and other pollutants. It also does a great job at cleaning the pavement, according to *nasweeper.com*. A lot of people don't think that littering is a big deal. I beg to differ, because if it keeps progressing and trash keeps flooding our good water, we won't have any more resources from the sea for food.

Amirahcoe Mason *is a sophomore at John O'Connell. He is a San Francisco native and plays varsity basketball after school. Amir enjoys role playing games and wants to learn more about the planets. He plans to go to college for business. If he could go anywhere in the world, he would go to Italy.*

WHERE WERE YA AT WHILE WE WAS IN GLOBAL WARMING?

In the case of global warming, some acknowledge the issue while some are aware and do nothing. Sea levels are rising and ice caps are melting because humans are putting CO_2 in the air, which weakens the atmosphere and exposes the earth to heat. *Rolling Stone* states, "A multi-meter sea level rise would become practically unavoidable." Everyone should care about this because on a small scale, coastal cities will be flooded. A consequence of this may be that the ground could be rendered useless or uninhabitable by way of flooding, land erosion, or water pollution.

Global warming, which has caused sea levels to rise, affects me because I live in a coastal city. Rising sea levels will cause coastal flooding in cities like San Francisco, Los Angeles, and San Diego. As stated on *EPA.gov*, "The impacts of climate change are likely to worsen many problems that coastal areas already face. Shoreline erosion, coastal flooding, and water pollution affect man-made infrastructure and coastal ecosystems." I feel that this may affect individuals that live in coastal areas greatly because all of the above situations are very plausible and may already be happening. For those who live on the coast, rising sea levels will be a problem.

Others areas in the world have been affected by global warming through rising sea levels. For example, as stated in an article in *Scientific American*,

"The tiny Pacific island of Kiribati was heavily affected, so much so that they declared the land uninhabitable." The Gulf Coast has been affected in a similar way, which is heavy flooding.

If we want to prepare for and prevent flooding in the future, some options we have are cutting carbon emissions. If we do this, it means the atmosphere will stop getting damage from foreign gases like carbon dioxide. We can use alternate fuel sources, like electricity. This method may have its own emissions, but they are not as dangerous as carbon dioxide molecules.

You can connect the sea levels rising directly to global warming because a graph on *climate.org* shows that sea levels have been rising since the 1800s. If you make the historical connection between air pollution (carbon dioxide) and global warming, you know that during this period, there was a huge surge in coal use known as the Industrial Revolution. This means that the air has been being poisoned for about 216 years.

Returning to the idea of preventing flooding, we have a couple of solutions—some realistic, some futuristic, and some already in service. One option is to build dams, an option that is already in use, which slows water down. Another option is making pits and tunnels to re-direct water to a place where it may be needed, for example, toward crops. Other options are flood barriers, especially the metal versions. According to the BBC, this kind is very mobile and inexpensive, making it standard issue in the United Kingdom.

In conclusion, I can gather from this evidence that global warming affects too many things to be ignored. One thing affected by global warming is our atmosphere, which is being weakened by greenhouse gases. Another thing affected by climate change was the entire island country of Kiribati; its land was rendered useless by flooding. We should invest in the idea of flood protection so that catastrophes and incidents like Hurricane Katrina and Superstorm Sandy can be easily survived.

Cristobal Cardenas *is fourteen and was born in San Francisco. He likes to play video games with his friends and listen to music. Cristobal's favorite subject in school is math, and he likes to solve equations in algebra. Though Cristobal enjoys playing soccer, his favorite sports to watch on TV are baseball and football, and he has high hopes for the Giants this season!*

FROM ICE TO WATER

The Antarctic ecosystem is being greatly affected by global warming. Biotic and abiotic factors that interact within the ecosystem include species, like walruses and penguins who depend on ice, and the ocean which surrounds Antarctica. In the larger global ecosystem, Antarctica is impacted by the heat caused by global warming, which is mostly caused by human impact. Humans release gases into the air, which affects the Antarctic ecosystem. The greenhouse gases that are released from Antarctica's ice create a cycle that heats up other ice and melts it, which also contributes to global warming. Antarctica is losing a lot of its terrain because of how much warmer it is getting. "Over 3,000 kilometers were lost from the Larsen B ice shelf (which is in the Eastern part of Antarctica) in just 35 days in February–March 2002, compared to 7,000 sq. kilometers for all ice shelves in the region over the last 50 years," according to *Science Daily*. The amount of ice lost in Antarctica ruins its geography, which hurts the whole ecosystem Antarctica has that depends on ice.

The West Antarctic ice sheet has been losing a bunch of its ice. Scientists at *climate.org* say, "The acceleration of ice loss from the West Antarctic Ice Sheet has doubled in recent years. Scientists estimate the loss of ice from the West Antarctic ice sheet to be from 47 to 148 cubic kilometers per year." The loss of the ice sheets contributes to another process, which has to do with the melted water from the ice. Snow sitting on top of the sea ice reflects about ninety percent of the sun's energy. This energy is then absorbed by the ocean. Since the water is absorbing the sun's energy, it becomes warmer. As the World Wildlife Fund explains: "As the open water of the ocean absorbs more heat and causes more sea ice to disappear, it exposes even more water and another feedback process has begun." This process continues because of the constant ice loss that happens in the ice sheets, which puts more and more water into the ocean and as a result, this keeps happening.

A result of ice loss is habitat loss. Animals like penguins depend on ice as a habitat. Because of ice loss, penguins lose their habitats. Since there's less ice, there are fewer resources, which means more competition between the species that live in the ecosystem. "In the Arctic a whole ecosystem relies on the presence of sea ice," writes the World Wildlife Fund. Sea ice and freshwater glacial ice are melting, and many species find it increasingly hard to adapt to the escalating changes. As a result of habitat loss, species can go extinct because of how many problems it leads to.

There are many species that are hurt by global warming, mainly because of the ice loss, which hurts the habitats and populations of species that depend on ice and snow—basically, every species that lives there. One penguin colony being affected by this is the Emperor penguin colony at Terra Adelie in Antarctica. According to the World Wildlife Fund, "The Emperor penguin colony could decrease by 95% before the end of the century, if the sea ice continues to decline at the current rate." This would place the population at a serious risk of extinction. In Antarctica, populations and habitats are all decreasing, which hurts the entire ecosystem of Antarctica. If this continues, there will no longer be as many animals as there were before.

People don't realize that they are contributing to global warming, which hurts not only their place, but places around the world. People add gases into the air. These gases go into the atmosphere, which makes it thicker, which means that it absorbs the sun's rays but doesn't release as much heat as it is supposed to, which makes the whole world warmer. When the whole world gets warmer, there is loss of ice in Antarctica, and a lot of changes to those ecosystems. I think that global warming is not really noticed by a lot of people, and it hurts a lot of ecosystems that are not around the people who are contributing to it. This affects me because I would not like to see animals go extinct. I think that penguins are cool animals, and don't want to see them become extinct. To decrease global warming, people will need to not release as much gas as they used to before, and not use as many things, like cars, that release gases into the air. Even though you're not in an ecosystem, you can still ruin it.

Jesus Saravia *was born in El Salvador in 1999. After a short stop in Daly City, he moved to San Francisco. He loves Kit-Kats and eats one almost every day. Jesus would like to be a politician and is currently involved in the Youth in Government program at the YMCA. He wishes he'd been in high school in what he thinks was a simpler time—the 1980s.*

OUR CHANGING NEIGHBORHOOD

According to the Anti-Eviction Mapping Project, more than 11,000 people have been evicted due to the Ellis Act, demolition, and owner-move-in in San Francisco between 1997 to 2013. This statistic is another example of the gentrification happening in many neighborhoods in San Francisco. I believe gentrification is negative in many ways, but can be seen as positive in other ways.

Gentrification leads to the eviction of people who don't have privilege. It helps the rich and hurts the poor. It can destroy social communities and cultures. It displaces people with low or medium incomes.

Many may question how gentrification is good. An example is that gentrification has decreased violence in the Mission District of San Francisco. A taxi driver once told me that back in the 1990s, the Mission District was filled with lots of gangs and violence. People couldn't go outside at night without the possibility of getting assaulted or robbed. Gentrification changed that. Now there are fewer gangs and the Mission District is safer.

The decrease of violence in the Mission District makes the neighborhood more attractive to new residents of different cultures and races and those with high incomes because they feel safer. Gentrification also led to the construction of new buildings and the remodeling of older ones. Streets and bus stops were also improved, making the neighborhood safer.

Not everything is a bed of roses, though. Yes, the Mission District's violence has decreased and many neighborhood buildings were remodeled, but not everyone benefits. Gentrification has disrupted the social equilibrium in the Mission. More than 11,000 people have been evicted in San Francisco. As I walk in the Mission District, I notice there are more restaurants catering to non-Hispanic customers. The prices of products in stores are getting more expensive and not everyone can afford them.

The displacement of people from their homes, some of whom have lived here for years, is unacceptable. When a new building is built or remodeled, the poor will often not be able to afford higher rent or will be evicted. Wealthier people will often move in. Families with children who no longer have a home are sometimes forced to sleep on the streets or in a car. They also face hunger. People shouldn't have to worry about whether or not they can pay rent or afford to eat.

The Ellis Act is the California law that allows landlords to evict tenants legally in order to go out of business, but many landlords do it with the purpose of building new buildings or remodeling. But many do not know that there are many types of evictions in San Francisco. According to the Anti-Eviction Mapping Project of San Francisco, the majority of evictions from 2010 to 2015 in the city were caused by breach, which means the failure to pay rent, totalling 2,816 evictions during that five-year period. Nuisance, or not following tenant rules, was the second leading cause of evictions from 2010 to 2015, equaling 1,627 evictions. It seems peculiar that there are more nuisance evictions than Ellis Act evictions. Owner move-in was the third highest cause of evictions from 2010 to 2015, totalling 1,058 evictions. Owner-move-in is an eviction strategy that is not talked a lot about in the media, but one should really consider owner-move-in as an important factor in gentrification since it has more eviction notices than the Ellis Act. Owner-move-in allows landlords to evict people living in apartments or houses if the landlord wants to move in. The Ellis Act came in at fourth place with 570 evictions from 2010 to 2015.

The city of San Francisco is similar to an ecosystem because we have both biotic and abiotic factors interacting. The Mission District is an ecosystem because we have humans and lots of resources. There is a lot of biodiversity in this ecosystem. We have Latinos, African Americans, Caucasians, Asians, and other races who benefit and don't benefit from the ecosystem's resources. We have landlords, tenants, politicians, citizens, workers, and others. The biotic factors in this ecosystem are humans who are competing for the resources. The abiotic factors in this ecosystem are houses, streets, apartments, money, laws, institutions, stores, parks, food, and transportation.

In this ecosystem we have reached our carrying capacity, in that the Mission only has so much housing. When a new tenant moves in, someone has to go and the former tenants may not have somewhere else to go in the neighborhood. The concept of parasitism applies in this ecosystem because the new tenants benefit at the expense of the former tenants. The way gentrification relates to parasitism is that the new tenants take the house or apartment that the former tenants have worked so hard for. The concept of competition in this ecosystem is that the new tenants and former tenants need housing, which is a limited resource. The concept of predation in this ecosystem applies because the former tenants are the prey while the new tenants are the predators. Because both parties need the resources and fight for it, the higher income tenants win and the former tenants lose. The new tenants are taking away a resource from the former tenants, leaving them with nothing.

How can we make gentrification less harmful? Well, we have to understand that gentrification is not an easy issue to deal with. However, there are many ways we can reduce the harmful effects of gentrification. Rent increases need to be better regulated. Evictions also need to be regulated. For example, the Ellis Act should be amended so it makes it more difficult for landlords to evict long-time residents. I believe that landlords have the right to make a profit, but they need to be considerate of all income levels. In San Francisco the local government has to find ways to have more affordable housing. One way to do that is by building new housing that would be fifty percent affordable and the other half free market. We also need to find a way for the contractors to make a profit from building these housing units.

We have to understand that the people whose neighborhoods are getting gentrified are sometimes not well educated. That means many evicted tenants are not aware of how to defend themselves. Also, a lot of evicted tenants are undocumented immigrants, and may fear stepping into a courtroom to fight their eviction. When a former tenant does file a civil suit, many times they end up losing their case. We have to be fair to everyone.

I hope one understands that gentrification is a good and bad thing. Gentrification can reduce violence and make neighborhoods nicer, but it can destroy communities and affect individuals with less privilege. I believe that we as citizens have the right to be heard and do something about this issue. We have to stand up and reduce the harmful effects of gentrification. It is our duty to defend those who need help and to make our ecosystems a nicer place to live. Gentrification, like racism, cannot be stopped. But we can reduce the harmful effects of gentrification. We can work together to have justice for those who are wrongly evicted. Will you do something to help your community?

IMAGINE HOW BEAUTIFUL
THE WORLD WOULD BE

Learning, growing, and taking responsibility

Lily Quijada-Madrid is *working toward becoming a pediatrician, the perfect pairing of her love of science and working with kids. When not playing soccer, reading mystery novels, or studying Greek mythology, Lily dreams of eating Nutella-filled crepes in Paris.*

THE NEW DIVERSITY OF HONORS

On the day I found out I was moving from Napa all the way to San Francisco, it was the worst thing I had ever heard my mom say. Imagine being at the most perfect school where you had the best grades, the most rockin' friends, and everything worked well—and then you had to move. It's the most horrible thought knowing your whole life may change, including who you spend time with. When I changed schools I no longer had honors classes, which really affected my first semester of tenth grade.

During the first day of my sophomore year, I was looking forward to what kinds of classes I would get on my schedule. But then, when I first received my schedule, I was quite surprised by what I saw. I was absolutely devastated that I had no honors or AP classes! Immediately I thought there must have been a mistake. I thought to myself, I already hate this school. How am I ever going to get noticed if I never even took an honors or AP class in my high school years? I figured that something was definitely wrong, and I should talk to a staff member, teacher, or counselor as soon as possible to fix it.

I turned to my right and saw someone who looked as if they could help me. I asked them, "Can you please tell me where I can find someone to help with different class selections to change my schedule?"

The teacher told me, "Since we are such a small school, we don't have a huge variety to choose from." I prayed to myself, *Please say that you have some type of honors or AP class to take*. But the teacher told me that there aren't any honors classes.

I felt very disappointed because I liked having good grades, an outstanding GPA that could be higher than a 4.0, and just being the best student that I could be. Who wouldn't want that? But I ignored it and told the teacher respectfully, "Thank you for all your help anyways." I believed it was best for me to see how it would go and if I liked it or not.

The first few weeks were an emotional rollercoaster for me. I noticed that in my math class I was already learning what I learned before, and in my science class I was basically learning the same things that I did freshman year. I thought to myself, *This isn't right*. I need to do something about this. I want a better challenge, and where I am at, it's not going to happen.

I went home that day and explained what I felt to my older sister, and she thought we should come up with a plan to fix my sticky situation. We thought for a second and that was when the idea hit me. I explained, "Maybe I should make a petition or something, and survey students who could possibly feel how I feel and ask them what we can do to fix this." She thought it was a great idea, and I made a vow to myself that the next day, I would go around and ask people if they think John O'Connell should have honors classes.

Once I recorded the data, I noticed that some people I surveyed agreed with my side of having honors, but others didn't. For example, my great friend Victoria Louie said, "Yes, we should have honors classes because everyone is at different levels of learning; some people learn quicker than others." However, I also noticed that a magnificent teacher of mine named Ms. Aguirre said, "It would be beneficial to some students, but it can create tracking, where 'higher' achieving students take all honors/AP together all day and then 'lower tracked' students in regular classes are together all day and the classes are lower quality."

Looking at the responses I received, both answers reminded me of parasitism. Parasitism is when one organism benefits while the other is harmed. If you think of a school as an ecosystem, you would see the relationship between the biotic and abiotic factors like the students, teachers, and the furniture in classrooms, including the chairs and tables and how they are organized. In this ecosystem, it seemed like some people could benefit from having honors, but then others could be hurt because it could make them feel bad about themselves by not experiencing the same high expectations as the ones we have for the "higher-tracked" students.

However, this situation is also an example of mutualism, because whether or not a student has higher expectations than others, all students will rise through their education no matter what learning level they're at. For example, if one student was in Honors English but another was in prep, it wouldn't matter that they are learning at different levels because both learn. This benefits the "smarter" student, but the other student benefits as well, because both will continue learning and expanding their minds even if they learn at different levels.

Originally I thought that having honors classes would benefit everybody and would be very mutualistic for teachers and students, because of the different ways teachers could teach that could be enjoyable for all. But as I looked through the survey, I realized that having honors classes put into my school would make my ecosystem die out, rather than making it more stable by adding what I want.

When you think of biodiversity, you think of having a variety of plants and animals in a certain ecosystem. Biodiversity is good because it supports an ecosystem's stability. When you think of this issue, you might think having a wider variety of classes in schools would benefit all students' grades. But does it really?

As I looked more into my decision of having honors, I realized not very many students would benefit from it. I was just thinking about what is best for me, and not my school. Adding honors classes would create enormous problems.

For example, it could create a very racist outcome. Asian students are more likely to apply for honors classes than black or Latino students, which is not what I want for my peers. Diane Rado from *The Tribune Report* noted, "The fact is that many times these 'gaps' are not about ability gaps. They start as morale gaps or gaps based on the misperception by the students or families that certain tracks are for certain kinds of students." She went on to say, "It's why we seem to rarely see high-achieving Latino students applying to our honors classes while we often have even low-achieving Asian students applying without any expectation of acceptance."

If anything, I want students to look at having honors classes as an advantage for their grades. But as it turns out, it's more of a disadvantage. It could lead to making black or Latino students not feel welcome in certain classes simply because there aren't as many of them in honors courses. This made me realize that the diversity in classrooms would change if my school decided to include honors courses. I definitely would not want a racist outcome to be the result of changing the classes at my school.

Having certain students in my ecosystem think differently about themselves not only could hurt them in this situation, but possibly in every aspect of life. As students grow, their mindset about wanting to have a successful future could change if they feel down because they aren't learning at the same level as other students. Having honors classes would not benefit all students and make the biodiversity in schools worse than it was. After thinking all this through, I've decided that I know what is best for my school now.

Biodiversity is an important factor for all ecosystems. Having a larger variety of classes to choose from by adding honors in schools will benefit several students, but I want to think of school as a mutual learning experience. I think it would be best if my school had a new system, where all students could take different classes where they could be challenged but not separate. I first felt that moving to San Francisco was going to be the worst thing that happened to my life, since before I was at the most perfect school. But as a new member of this ecosystem, I just want the best possible outcome, and to make everyone feel both welcome and challenged.

Gabriela Martinez *was born in San Francisco. She enjoys writing poems and short fiction. Music helps her when she has writer's block because she is able to find hidden meanings and inspiration in the lyrics. Gabby doesn't want to grow up because she doesn't want to have to choose only one passion (but maybe she doesn't have to!)*

JUST KEEP SWIMMING

You're going to the beach with your family. You have a great day planned ahead of you: a sunny day, a beautiful blue beach with the glint of the sun reflecting perfectly on the water, making it as clear as the cloudless sky. The sand will feel warm, almost hot even. You'll put on your swimsuit and go for a swim, and it will be as if you are floating on thin air, floating for yards and miles, until the waves drop you back into the sand. After, you'll go back to your family to enjoy the nice picnic you have planned: some sandwiches, some fruit, the usual family beach basket. It's what you took last time. You pack everything into the car: picnic basket, blankets, towels, and surfboard.

Finally, you arrive at the beach. You step out and take a look at the scenery. The sea isn't clear. It's as if ink spilled all over to replace the water. You're stepping on something. You remove your foot to see a plastic water bottle under the sand, filled with, yep, sand. Your swimsuit is on; you might as well test those murky waters. You walk to the ocean, but see that even more trash litters the ground. Colorful speckles scatter on the ocean surface. The speckles are soda cans, more bottles, wrappers, and pieces of a broken surfboard. You get into the water and waddle out into the ocean. The water feels heavy as your swimsuit starts to get stained, and you start to realize the water is more oil than ocean.

"This doesn't feel like I'm floating," you say to yourself. You see that what you mistook for foam in the waves was actually even more trash. You wonder if there is an end to all of this litter. You lose your appetite. You should've listened to your relatives and gone to the country club. At least there they have people to pick up the trash.

Human actions are destroying San Francisco Bay Area beaches. Three of the dirtiest beaches in the state of California are in the Bay Area, and this is due to pollution and not paying attention to where we put our waste. If we want to see clean, clear ocean waters again, we need to change how we treat our Bay Area beaches.

Heal the Bay created their most recent annual Beach Bummer List of the ten most polluted beaches in California in 2015. Three of the beaches on the list were San Francisco's Sunnydale Cove, San Mateo's Marina Lagoon, and Santa Cruz's Cowell Beach. These beaches made the list because of their weekly high levels of bacterial pollution. People who came into contact with these polluted beaches might have gotten ear infections, stomach flu, upper respiratory infections, or skin rashes.

Bay Area beaches form an ecosystem because biotic factors such as fish, plants, crustaceans, and mammals and abiotic factors such as the water and sand form a community that interact with each other and their environment. We humans are also a part of this ecosystem because we interact with the ocean by fishing, swimming in the water, and even using it for travel. The existence of many different kinds of plants and animals in the Bay Area beaches makes this ecosystem an example of biodiversity. We have not been treating the beaches and their ecosystems with the care we should be. Pollution is a common way we are harming the ocean's ecosystem.

Trash and sewage are two types of pollution problems that Bay Area beaches are facing. We either don't know or don't care about how to dispose of trash properly, so we litter our trash in the streets. The trash is then washed out by the rain and then eventually ends up at the ocean shores and in the water. Other times trash is intentionally dumped by boats into the water, using the water as a big trash can. Trash is harmful to the ocean's ecosystem because the fish eat it and die. According to Save The Bay, "Trash collects in thousands of Bay Area creeks and storm drains, which flow to the Bay. Every year, marine wildlife like harbor seals, sea birds and turtles, are killed when they eat or are entangled in trash." One of San Francisco Bay Keeper's scientists, Ian Wren, said the beach pollution might be because of storm water runoff and the city's faulty sewer layout, which causes harmful bacterial waste to be dumped into the ocean. He said, "San Francisco, like many other cities on the bay, dumps its sewage into the Bay, adding bacteria to the bay water."

There was a time when I was in second grade on a field trip to the beach with my class, and, like any little kid, I was excited to go. I always loved the beach, the salty yet sweet air filling my lungs and the water feeling like a cold welcome from Poseidon. When I got there, I immediately took off my shoes and stuck my feet into the sand, yet I couldn't take it and ran into the ocean. But as I was running, I tripped on a water bottle full of sand and fell face first just as the tide was coming in. As expected, I got wet from the shoulder up. When I got up and looked into the water, I saw a can of Dr. Pepper washed up with the tide. After seeing that, I was disgusted. How can someone taint something so beautiful? I decided to go back to the sand. I'd had enough of the ocean for the day.

Ever since that day, I haven't looked at the ocean the same way. Every time I see it I think, *what other dirty secrets are you hiding*? I've thought about it so much that the ocean has lost most of its splendor, and I've admired it less and less, to the point where I'm scared to go in it because of the litter I might find. Yet I still go there whenever something upsetting happens to me, because I still hold respect for the ocean. I respect it because no matter how much pain we inflict upon it, it is still there to provide us with its food, its experience, its secrets, and its inspiration. That's why I am writing this, so that I may not be scared anymore to venture into the ocean's depths, and to return the sea to its original state of regality.

The Bay Area beaches can still be saved. If we take precautions and take it upon ourselves to fix the contamination problem in the Bay Area beaches, then we can prevent a tragedy of the commons in which all of the world's oceans are destroyed because individuals were acting in their own best interests instead of thinking about the success of the entire ocean's ecosystem. There are several ways we can avoid the human destruction of the ecosystems in the Bay Area beaches. For a start, we can use less plastic and recycle the plastic we do use since it never decomposes, and it ends up either in open water as litter or in a seal's stomach causing it to die. Also, we can dispose of our waste properly by recycling, composting, and putting the rest of the garbage in the trash bins and not in the streets. We can also volunteer with various organizations to help clean the ocean and its beaches or simply raise awareness of ocean importance within our neighborhoods.

Now picture this: people raising awareness of ocean pollution and volunteers cleaning the beaches with smiles on their faces. After a while there won't be any more trash on the beaches, no more soda cans floating on the shores, no more oil spills, a well-functioning sewage system that disposes of waste properly, and your family finally having the perfect picnic, that perfect day, next to the white sandy beaches and the clear blue ocean.

Gary Ray *is a fifteen-year-old Bay Area native. He can often be found on the basketball court. He likes to play video games, read manga, and hang out with his two dogs. In the future, Gary hopes to graduate from college with a degree in a tech-related field and then start his own video game company. If he could live anywhere in the world, he would live in Canada.*

BURNING TORNADOS

I'm Gary Ray and I live in Vallejo, California. One day I was watching the news, and they were talking about fire tornados in San Jose. I was very curious about this because my grandma lives down there. A few months later, I had to write an essay about something and thought this was the perfect time to learn more about the fire tornados and wildfires. The three things I'm going to be talking about are the California drought's effect on wildfires and how can we reduce that, how people cause most wildfires, and lastly, how controlled burns are the best way to help reduce the number of wildfires.

Because of the drought in California, we have seen an average of 1,600 more wildfires in California a year from 2011 through 2015. Due to this drought, the land in California is very dry. This makes it easier to spark a wildfire, especially in Southern California. For example, the wildfires in San Jose that were spinning in circles (as if there was a fire inside of a tornado) were caused because of dry plants and dry land, which are very flammable.

Droughts are a part of nature and always have been. To be honest, as of right now, there is no way to stop a drought. There is a way to reduce the effect of the drought, which can reduce the additional wildfires caused by the dry land. The Sacramento Suburban Water District (SSWD) has been required to reduce its water use by thirty-two percent in order to help reduce the effect of the drought. Farmers are using recycled water, which is when they reuse dirty water later by filtering it.

One thing that was shocking to me when I first started to research wildfires was that ninety percent of wildfires are caused by humans. What makes this shocking is that many of these fires are caused by gardening tools, such as WeedWackers and lawn mowers. The reasons why these two gardening tools cause wildfires is because people do not use these tools in the correct way. With a WeedWacker you need to use a plastic blade instead of a metal blade because the metal blade can spark a fire if it hits a rock, especially in dry grass.

A lot of people don't know that you are not supposed to use lawn mowers on dry grass or weeds because unlike WeedWhackers, lawn mowers are supposed to be used on the lawn only. This is why they don't need plastic blades and only come with metal blades, which can spark fires against a small rock. Don't use lawn mowers or WeedWackers while it is too windy outside. People can learn how to use these tools properly. They should also have a cellular device on them at all times while they're landscaping, and a fire extinguisher. I believe that people should be aware of their surroundings at all times.

Controlled fires and controlled burns are two ways that firefighters prevent a big wildfire from happening. The way they do this is by burning vegetation off the forest floors. In other words, they start small fires to burn dead plants and some live ones as well to keep even more plants, animals, and humans safe and to keep the big fires from polluting the air even more. How wildfires affect the air depends on many factors, including weather conditions such as wind and humidity. I'm interested in the different ways wildfires affect both humans and animals.

I know that wildfires can cause animals to migrate into territory in which they do not belong, and then it happens again and again until you have a chain reaction. For example, I once saw on TV that in Australia, feral cats moved into the ecosystem of another group of animals' and kept killing all the animals that were injured by the wildfires. This was messing up the food chain and causing other animals to leave their ecosystem, which was putting both them and maybe other animals in danger. By that I mean they were basically wandering into new and unknown territory, making them easy targets.

I think controlled burns sound like the best method for protecting against wildfires. As I said before, there is no stopping wildfires, but there are ways to reduce the amount and the size of them. This is exactly what controlled burns are trying to do. I know some people may think that if people know how to use gardening tools properly that will mainly reduce the amount of wildfires, but I believe that controlled burns will do the most to help with both human-caused wildfires and natural wildfires. Face it, humans have always made mistakes and sometimes just teaching them won't help fully prevent our mistakes. What do you believe can help reduce the amount of wildfires after reading this?

August Bennett, *who goes by Gus, is a San Francisco native living in the Mission District. He is passionate about the art of photography and his favorite subject is math. He has visited China as an exchange student, and hopes to visit Japan in the near future.*

EBOLA: WESTERN AFRICA'S QUEEN OF SPADES

Ebola was in the news so much in 2014 that I can't remember the time I first heard of it. I was scared of it because of everything I had heard it could do to people. I heard that it had spread through Western Africa like wildfire, and that it was something that could threaten the safety of the rest of the world. I thought that the things it could do were scary, and wondered if it could spread outside Africa and turn into something that would be the end of humanity. Ebola has had a negative impact on the ecosystem of Western Africa because it has put the world on higher alert, killed thousands, and diminished social communities worldwide.

The 2014 Ebola outbreak is believed to have started from a single transmission from a fruit bat. It can't be traced to a single source for certain, but this explanation is generally agreed upon. Western Africa is an ecosystem because it has biotic factors, like people and bats, and also several abiotic factors as well, such as small homes and even the economy. The Baylor College of Medicine states, "[A] child is thought to have played in a tree with Ebola infected bats, so that he likely came in direct contact with the bats or their droppings. The virus transmitted by these bats is closely related to the Zaire Ebola virus." It's useful to know how the virus could have begun to spread, because it helps to track where it spreads and gives background to Ebola and the 2014 outbreak.

The virus has cost the world billions of dollars in treatments because of the equipment costs and the cost of hiring people who are willing to come into contact with people who have contracted the deadly disease. The World Bank has said, "The Bank Group estimates that these three countries will lose at least $1.6 billion in forgone economic growth in 2015 as a result of the epidemic." The countries have suffered devastating economic losses due to Ebola, but if it weren't for those losses, the Ebola outbreak could still be happening today.

Ebola is still around today, but the number of infected people has greatly decreased since 2014. Several weeks ago, the British Broadcasting Corporation released the following statement: "The World Health Organization declared West Africa Ebola-free last week, after all the affected countries had gone 42 days without a case. But then, just hours later, a death in Sierra Leone was confirmed to be from Ebola." This quote shows that Ebola is still around, but not nearly at as large a scale as it was when it spread during the outbreak in 2014.

Ebola is a form of parasitism because it uses humans as a host in order to spread worldwide, even if it means killing the human. Ebola could also be recognized as an invasive species because it doesn't belong in the ecosystem it exists in, and is causing trouble that wouldn't exist without it. The Ebola doctors are a form of altruism because they are willing to risk their own safety to treat others, even if they know that they could contract it.

In conclusion, Ebola has affected Western Africa because it has spread from bats to people, cost billions of dollars, and another outbreak could start at any time. Bats were mainly responsible for the 2014 outbreak. It also still exists today, but only from person-to-person transmission, not as the epidemic it was in 2014. Ebola is a deadly disease, and another outbreak could start at any time. If another Ebola outbreak starts, it would be important for several reasons. It could cause more trouble than it already has, and could further shrink the ecosystem of Western Africa.

Monique Partee *is a sophomore at John O'Connell High School. She is the youngest of three sisters and two brothers and loves her dog, a chi-weiner named Kiyah. She hopes to study forensic science in college before attending the police academy. Monique dreams of becoming a trusted officer, who protects the people without abusing her power. She is currently studying for her driver's license test as she turns sixteen in June.*

CALIFORNIANS STUCK IN A BAD HABIT

What is a drought? Why do droughts happen? Can we stop them? A drought is a period of below-average precipitation in a region. Droughts are caused by many factors such as climate change, natural variations in weather, and poorly planned infrastructure. Despite these causes, we as humans aren't making it any better. We're actually making it worse! We carelessly waste water and are ignorant about our natural ecosystem and how to appropriately adapt to it. Human behavior is making the California drought worse, but this can begin to change if people become aware of their influence on the ecosystem and their responsibility to conserve water.

California has been in the worst drought in recent history for four years. A drought is a long period of time with low or no precipitation. California has a lot of different types of ecosystems such as deserts, Mediterranean regions, and mountains. Although parts of California receive plenty of precipitation, there are many that don't. Regardless, all have been affected by the drought. People have caused a lot of these problems, making the drought worse by overusing water. Living organisms, such as humans and animals, are the biotic factors in California's ecosystem. Abiotic factors, such as soil and water, are being abused by the biotic factors, specifically humans. Our abuse of water is one of the reasons why we're in a drought. Naturally, California is not meant to sustain all these humans reproducing and living in the region. Because the population is so over its carrying capacity, we're slowly damaging the region. According to Nobel Prize-winning scientist Paul Crutzen, "Humans have caused such a significant environmental change that we have entered a new era known as the Anthropocene."

According to the United States Census Bureau, "California had a population of 38.8 million people in 2015." According to my estimates, the average person needs about 200 liters of water per day for personal use. That means that for personal consumption, we need about 7.8 billion liters of water per day in our state. This doesn't include water used by manufacturing and agriculture. Last I checked, it hasn't been raining billions of liters of water every day. It's probably safe to say that we have reached our carrying capacity in regard to water here in California, yet we continue to waste water.

Ways that we abuse water include, but are not limited to, over-showering, watering public and private green spaces, and agriculture. There are too many of us in California to share one resource. This is an example of a tragedy of the commons, where individuals are acting in their own best interest, but everyone is hurt because we are using up our resources. The outcome is that we are consuming our resources much faster than we should be. Humans have the urge of wanting more. People don't understand that they're not the only person and they're part of a bigger picture. For example, people are over-showering. We should be taking the advised five-minute showers, but instead *some* people are taking thirty-minute showers (*cough! cough!*). Another way people waste water is by letting the faucet run when cooking, or while they're brushing their teeth, or doing their make-up. This reflects people not understanding our region and its history. Knowing that we had a "megadrought" that lasted ten to twenty years during the medieval period, according to data from tree rings, we could learn or start conserving water earlier.

For those of us living here, we need to understand the drought before we can conserve water. California's infrastructure is set up to rely on snow. For example, people think El Niño will relieve the drought, but we need snow, and El Niño doesn't appear to be giving us that. Many people don't understand that snowmelt from the Sierra Nevada mountain range is brought to one of our many reservoirs via aqueducts. This is just one example of what people don't understand. Just because we're in an El Niño year doesn't mean we can start abusing water. We need to be educated on water use and conserving water so we don't use up all of our resources. Lack of conservation has led to the restrictions on water use. If we learned that earlier in our education, then maybe we wouldn't be as deep in the drought as we are now. It's not all our fault. The education system should have made this a bigger part of education. Instead of learning how to water paint in kindergarten, I could have been taught to conserve water. Then maybe I wouldn't be stuck in such a bad habit.

The most important thing is that you are not going to care about something if you don't know about it. We need to learn about our region, and what is going on, in addition to learning about something happening 3,000 miles away. The people within the state of California need to realize that this is a personal problem that affects everyone as individuals. Information is the most efficient way that human behavior will change, and this is why the education system needs to include the drought. The ads that I see at the bus stop only say "Conserve Water," they should have more information on them. They give so little information on what to do, how to do it, and why we have to do it. These ads only have a tiny website at the bottom and do not let people know that they are actually a part of the problem. My experience with the drought is not knowing what a drought was until 2012 when it was announced on the news that we were in a drought, and that we were going to have to start conserving water. I didn't know what a drought meant; I just knew that we weren't getting enough rain.

The information on the drought was confusing. I finally knew that the drought was serious when restaurants stopped serving water unless one asked for it, but that wasn't enough to make me want to conserve water. I didn't understand how I, as an individual, could affect the amount of water being saved somewhere, but now I do! We can't stop the drought, but we need to take responsibility and conserve water.

Karen Villanueva *is fourteen years old. She was born and raised in San Francisco, even though she lives in Oakland. She loves to eat, sleep, do homework (she finished the year with good grades), go biking in the park, and spend time with her three dogs. In the future, she plans to become a doctor or a vet in addition to being a secret spy.*

BEAUTIFUL, FRESH, AND CLEAN

When I first learned about pollution I was in the third grade. At first I didn't really have any thoughts about it, but when I started to learn more and more about it, I started to realize that pollution is important. Pollution is important because it causes many problems in the world today. Pollutants can come in many different forms and are found in water and air. Most of the world has polluted air. Pollution can make cities look grimy. In San Francisco, pollution looks like haze. San Francisco has pollution, too, because of the cars and factories. Pollution impacts people in San Francisco because it causes health problems. We can stop pollution.

San Francisco is mostly foggy, but sometimes it isn't fog. Sometimes it is haze. San Francisco is an ecosystem because abiotic and biotic factors interact with each other. Fog comes over the people, buildings, and water. Fog is an abiotic factor. Fog affects the biotic factors because it affects the people in everyday life. Fog affects people because it affects their outfits and plans. People, which are biotic factors, and cars, which are abiotic factors, interact with each other as well. San Francisco is a busy city and people drive cars for that reason. People work in factories which are abiotic factors, which also cause pollution.

Places in San Francisco including the Bayview, Western Addition, Downtown, and Potrero have lots of pollution. Pollution is a big problem in San Francisco because it gives us bad air and health problems. We can fix this problem.

Pollution, which is harmful substances that are bad for the air, is primarily caused by cars and factories. Cars cause the most pollution. Cars are everywhere.

According to *Biology Online*, "Pollution is the change in the environment caused by natural or artificial input of harmful contaminants into the environment." This may cause instability, disruption or harmful effects to the ecosystem.

We have pollution because people rely on things that pollute, like cars, airplanes, and factories. We rely on these things because it is convenient and easy. Even though most people know that pollution is bad, they still use their cars. This affects everyone and the environment. This is the tragedy of the commons because everyone is acting in their own best interest, but eventually everyone gets hurt. Clean and fresh air is a limited resource because of pollution. Pollution is an example of the tragedy of the commons because many people would rather take the car and not have clean and fresh air.

Pollution impacts people because they breathe in the bad air. When people breathe in the bad air it causes health problems like asthma, cancer, and loss of lung capacity. Fresh air is a limited resource. This is a problem because people need fresh air to survive. Not every neighborhood has fresh air. Some people in San Francisco don't have enough good air to breathe. They have to live near highways and factories since there are too many people in San Francisco. The city is at carrying capacity. The people in places like Bayview, Downtown, Western Addition, and Potrero are affected because of the pollution, which causes health problems. The people who live in that area may not breathe very well.

One way to solve this problem would be to plant more trees because trees give fresh air to all the people and could be good for the ozone layer. Other solutions to stop pollution would be to stop taking cars with only one person. Start carpooling or start taking trains and that will help lessen the pollution.

The future without pollution would look so fresh. What I mean by fresh is that everyone would have clean air to breathe and a lot more trees would grow. The trees would look green, have lots of leaves, and give us air; when it's breezy the trees would shake and their leaves would rustle in the wind. The sound is very soothing. The future would have clean air. It would look fresh, clear, and have blue skies without haze. In the future, there would be fewer cars and more trains. That means fewer car engine sounds, honks, and exhaust fumes. Also, there would be fewer health problems. People would be better and happier.

Pollution in San Francisco is a big problem. Pollution impacts people in San Francisco by causing health problems like cancer and asthma. We can change our behavior to stop pollution. Our city will look better in the future. I guarantee it. What are you going to do to stop pollution and to have a better future?

Victor Louie *is a very energetic and positive person, which he says he gets from his older sister. Victor is a shooting guard for the basketball team at John O'Connell High School. Most of his family, including his dad and uncle, are mechanics, which is a career he is considering for the future, alongside studying at a UC college and looking more closely into architecture. Victor has a great sense of humor and is always up for a good laugh.*

EDUCATION + CHANGE = R.I.P. DROUGHT

I used to have a peach tree that grew fresh, sweet, pink peaches, but it doesn't grow anything anymore. When I was younger, my sister, grandpa, and I used to water all the plants in my garden every week. But now, because of the California drought, it's not really an option and we can't water them anymore. The statistics of the drought make me worried about the future of California. I care because I can't do things that I normally would. I can't go to my favorite water parks like Raging Waters, and I have to take shorter showers (and I love my showers). My perspective is that I see my family members trying to make a change, and the students in school just ignore it and say that the drought doesn't matter and doesn't affect their lives. I want people to know that the drought is not going to end any time soon and the drought was not caused naturally. It was our fault.

California is an ecosystem with a lot of wildlife like plants, animals, water, etc. and a lot of biodiversity. There are many biotic and abiotic factors that exist in California's climate, but the drought is causing the animals to migrate to different places because they were used to the normal California climate. The drought has caused relationships like competition, parasitism, and the tragedy of the commons. The drought has caused competition between people and animals, because they are fighting for the limited drinking water. The government is using tax money the people paid and spending it on farms, which is parasitism. This is parasitism because people are losing but they are not benefitting and the farms are getting water from the government, which takes the money. People who are not educated about the drought are using more water to benefit themselves, water which could be used to refill reservoirs and provide more water to everyone. This is an example of the tragedy of the commons.

Under Governor Jerry Brown's administration, the California Legislature passed a sweeping groundwater law. The groundwater law makes it so that humans, "can't pump more water from underwater aquifers then can refill those aquifers" (*The New York Times*). People are using recycled water to water their plants. They are also doing less laundry, taking fewer showers, and not watering their lawns. When people are doing things that conserve water, it benefits other people and it helps with the drought problem; it is mutualism. But there are still some people who just ignore it.

I am thinking about taking action by trying to spread the news in small neighborhoods. I would want them to know about food prices and show them some past pictures of lakes that used to be filled with water and now look like dried up, empty, two-drops-left lakes. This would be an effective strategy because people don't pay attention much to information they hear. They want to see what is happening in person. If we don't take action now and decide to just talk about it, it will just be ignored. Actions speak louder than words. The drought is affecting everyday life for everyone. People are spending more money to buy fruits and vegetables from the market. People have to make changes in their lives to conserve water. But some people who don't have access to news don't know about the drought and what changes they need to make.

California used to be filled with water. Plants were healthy, the wildlife was okay, and we had a good water supply. But now that we are in the fifth year of a severe drought, everything seems to scare people. America is the third fastest growing country in the world and about one million immigrants come to California annually. Our resources can barely support everyone. This is exceeding California's carrying capacity and with the drought there is barely going to be enough water for everyone.

People are afraid that the drought might get in the way of their cultural traditions. I see all of my neighbors having gardens, and those gardens have been in their lives for a long time and have a lot of value to them. California produces nearly half of the country's fruits and vegetables. The drought caused the prices of fruit to increase by six percent and vegetables by three percent. About 17,000 farmers lost their jobs in 2014 and stand to lose $810 million if the drought keeps on going. The drought has caused the agricultural sector to pay $2.2 billion.

The drought is also endangering wildlife. There have been more sightings of California black bears and coyotes in cities. Scientists say that the drought is causing animals to move around because there is barely enough water for them to drink in their environment. Also, there have been 3,400 wildfires as of July 19, 2015, a thirty-six percent increase over a five-year average. The plants are dying in the dryness of the drought, and that is causing many worries for farmers.

Educating people with statistics like these is important and it would be dumb for anyone to ignore them. Instead of making laws about the drought that no one will follow, the government should educate everyone about the drought. If they educate people, other people will know about this problem and what they should do to help. Also, the educated can convince other people to help with the drought.

As humans, we do things that destroy ecosystems or damage them. Then the pollution caused from the destruction goes into the air and makes it unhealthy. For example, we destroy many forests to build factories and buildings. This process causes air pollution and releases carbon emissions into the atmosphere, which makes the weather warmer. Carbon dioxide traps heat in the middle of the atmosphere. This warm air which is higher in the atmosphere, tends to prevent the rising air motions that create thunderstorms and rainfall.

The California drought affects everyone's lives despite people acknowledging this problem. Food prices have increased and millions of dollars are being spent to help the drought crisis. We need to make a change. It's a must! I know it's going to take a lot of work and change to help end the drought, but it's worth it. To help save California from the drought we, the people, should make the little changes in our lives that will still help a lot. We can take shorter showers, use recycled water to water our lawns, and, most importantly, educate people about the drought and what we need to do to get past this disaster. These changes won't fix the drought right away, but if we all pitch in and help, they could do a lot for California.

Ruby Aguilar *is fifteen years old and was raised in San Francisco. She is a freshman at John O'Connell High School, where her favorite subject is science. She likes discovering new things and asking big questions. Outside of the classroom she can be found making art, tickling her little cousins, and eating her mom's homemade tortillas and pupusas.*

IT'S ALL ABOUT LITTER

People do what they feel like. For example, people litter. Then the trash goes into parks, on the sidewalks, and even into the ocean. This is a case of the tragedy of the commons, and it is a problem that involves us. People do things in their own way as if nothing bad is going to happen, but in the end, it is going to affect all of us. That relates to littering because some people just do it knowing it's going to make a problem in our community, but since it's not affecting them directly or at the moment, they keep doing it. I think part of the solution for this problem is educating people about what is happening and what could happen if they litter.

I interviewed five of my friends and they all said that they litter every day, probably more than two times a day. One of my peeps said that he litters thirteen times a day because, "YOLO" (you only live once). And when I asked, "Do you think that will affect us someday?" he said, "Hell no." This relates to the tragedy of the commons because he is littering because he thinks it's not affecting him and he doesn't know what the effect will be. We need people to have more information, not just about the city, but about the ocean as well.

An ecosystem has biotic and abiotic factors. Biotic factors are living things, like sea animals, and abiotic factors are nonliving thing, like rocks and water. This information shows that the ocean is an ecosystem. We're also connected to it because we eat fish and crabs and things like that.

What I've found on the Internet is very interesting. I have found information that tells me how our garbage can reach the ocean. I found out that there is trash twice the size of Texas in the Pacific Ocean. If garbage is not disposed of correctly, it can reach the sea and also might reach the sea shore. I feel like this is a problem because it affects our health, sea animals, and tourism. People used to think that throwing the garbage in the sea would get rid of it, but the trash actually just gets broken into smaller pieces.

I feel like people need to be educated more about this. What I mean by that is that teachers need to teach more about littering and what littering causes. I also think we all need to learn more about what an ecosystem is, because there's life in the ocean and we throw trash in there, and it's affecting not only the animals, but also us. I decided to write about this topic because I found it interesting how we know what might be happening, yet don't stop doing the thing that's causing the problem. Every time I go out I see littering happen, and it makes me feel confused because there might be a trash can nearby, but some people refuse to walk and throw trash away.

Ayiana Escobar *is fifteen years old and loves playing softball for the John O'Connell High School team. She has siblings who love her and whom she adores, and is also the proud owner of a cat who is appropriately named Fatty. She hopes you enjoy her essay.*

THE CRAZY FIVE-YEAR DROUGHT

I saw the impact of the California drought through two pictures of a lake. In 2011 it looked like a beautiful lake filled up with water, and now, in 2016, it just looks like a dry river with nasty water. Humans are impacting the ecosystem in this way by being wasteful and not caring enough about whether we die or live. One ecosystem affected by this human impact is California. In California there are biotic factors, which are humans, animals, and plants, and water, which is an abiotic factor. Humans are wasting water. California's drought is getting out of control because the governor is not helping enough, and also some people just ignore the drought. We have to stop playing and help out.

California's driest years on record were 2014 and 2015. This means the drought is getting worse, and we have to do something about it. The governor, Jerry Brown, declared a state of emergency because California is in its fifth year of our drought and it's not getting better. These dry years snuck up on people, and some families and people were not prepared for the drought.

People in the government need to speak out and tell people to use water carefully. I learned about the drought in school, and it really scared me. So I thought, *Why not talk about it and help out the people of California?*

People can be quite selfish with water, and that's just like the tragedy of the commons—eventually everyone's going to get hurt because no one is helping out anyone and they're being selfish with water. The ecosystem of California is past its carrying capacity because we keep getting more people and less water. We have to find a way for those factors to be equal. Since we can't change the population, we have to change amount of water we have, and the way we do that is by using less.

I chose to write about this topic because people need to understand that this isn't a joke and we need to help out the environment and do something different. It matters to me because I have a family that lives in poverty, and if we don't have enough water, what will we do? It affects my life because I love my family. I would sacrifice everything for my family. It will affect those of us in poverty more than others because we are struggling with rent and other stuff, and we can't pay more for water. Something that has changed for my family since the drought began is that we are all worried now. We use water differently now. We wash the dishes with just a little bit of water, we all have water bottles, and we try to work together to save water.

I surveyed my classmates and this is what they thought about the drought: they felt like the drought is "scary," "really bad," "crazy," "sad," and "no different." What they mean by "no different" is they don't see anything changing and they think the drought is fake. Six out of seven students who I asked about the drought thought it was really bad. They said that it affected them by making them use less water, save water, and stop watering their plants. Two out of seven students said it didn't affect them at all.

Students said that the drought has made them feel "conservative," "scared," and "worried because some people don't care." Two out of seven students said that it doesn't make them feel anything.

I asked, "Do you like the drought, yes or no?" Two students said yes. One of them said, "I like the drought because now people will start developing methods to save." Five students said no. One of them said, "We will have less water, we won't have a lot of clean water, water is limited and it's a competition in California. "Then I asked, "Are you scared of the drought?" I got four yeses from students. One student said, "What if we don't have water? We all die. Competition can really bring out violence and cost lots of money, and money is really tight for California." I also got three nos to this question. This is what they said: "It's really bad, but we still have water and nothing has changed."

I disagree with the students who said we still have water and nothing has changed. I looked more into it and we don't have a lot of water left. What I agree with is that competition can bring out violence and cost lots of money, and money's really tight for California. Everyone has to help each other.

Humans could change their behavior by doing these suggestions from my classmates: change your shower time, turn off the faucet while brushing your teeth, and only use the water that you need. There should be no more pools and fewer gardens. Other ways you can save us from the drought or help out is to do what my family and I do. We use water carefully and we run the faucet for food only, or for dishes. We don't use the dishwasher anymore. If the drought is ever going to be over, we need to follow these tips. Save more water so we won't be in a drought anymore.

Javier Flores *is a sophomore at John O'Connell High School. He enjoys sports, horror movies, and spending time by the San Francisco Bay. If Javier could go anywhere he would go to New York City. He has mixed emotions about writing and would like to know more about current events.*

SMOG CITY

Do you believe in global warming? Many people don't, and there are many websites that are dedicated to calling global warming a myth. It is surprising to find so many articles by nonbelievers while scrolling through the Internet. A lot of people have heard of the phrase global warming, but few know its true meaning. NASA defines global warming as "the average global surface temperature increase from human emissions." This means that the current climate change is the result of humans using natural resources. People should take global warming more seriously and every person has a role to play to help take action against global warming.

There are many factors that go into global warming. A city can have negative effects on the ecosystem of the natural world when certain abiotic and biotic factors work together. Abiotic factors are nonliving organisms that affect an ecosystem, and biotic factors are living organisms that affect ecosystems. Some abiotic factors are fossil fuels, machines and factories. Some biotic factors are humans, animals and the living things in the oceans.

Humans commit parasitism, which is when one member benefits at the expense of the other. This is true in the case of coral reefs. *The Guardian* states, "Warming causes mass mortality of warm-water corals through bleaching as well as through biotic diseases, resulting in declines in coral abundance and biodiversity." This quote means that biotic diseases and declines in coral abundance and biodiversity are all caused by global warming. If there were less bleaching, there would be more homes for fish. If there were more homes for fish, that would mean there would be more biodiversity in the ecosystem. This would mean that the ecosystem that the fish live in would become more stable with all the variety of species of fish.

This is not just is happening in one area. "A report showing climate disruption is already leaving deep imprints on every sector of the environment," *Scientific American* states. What this says is that global warming is a serious problem and doesn't just happen in one area. It is a problem that is occurring all over the world. For example, *Scientific American* also states that the "Caribbean and Southeast will see increases in wind, rain and storm surges. California and the Southwest will see drier summers. All will see impacts to human health, water supply, agriculture and other aspects of society." Still, some people are not taking global warming seriously.

Other examples of serious effects caused by global warming have occurred through droughts, floods, and other extreme natural disasters. Many of these events have been caused by humans. We are only hurting ourselves. NASA states, "In 2015, there were 10 weather and climate disaster events in the U.S. each with losses exceeding $1 billion...Overall, these resulted in the deaths of 155 people and had significant economic effects." This is concrete evidence that shows how we are hurting each other. Also this shows us how many people died because of careless things. We could have done so much to prevent this, and we can do more to keep it from getting worse.

All of this is caused by humans. "Most climate scientists agree the main cause of the current global warming trend is human expansion of the greenhouse effect." This statement published by NASA states that all of the current heat trends are almost definitely caused by humans. This lets us know, more or less, that this is all our fault and that we should fix these problems. Some ways we can do this are by biking, using public transportation, and spreading awareness. Biking and using public transportation would help because they minimize the amounts of fossil fuels we use and help make our carbon footprint smaller. Spreading awareness is important because once someone recognizes a problem they will want to help create a solution. These are examples of altruism. However, people are still stuck in the tragedy of the commons, people could do little things to help out global warming but would rather take the easy way out. An example of this is San Francisco's Super Bowl City. People could have walked, biked, or taken the bus, but instead they decided to take their own cars to downtown San Francisco. That caused lots of traffic and everyone was stuck in it. These little things pile up, leading to tons of pollution.

In conclusion, global warming is affecting coral reefs in the ocean, the Caribbean, and also is causing droughts in the West. There have been many lives lost. This physical evidence shows us that we need to take global warming more seriously. Once we realize it is more threatening than we thought, then we can make more sacrifices. If we gave up more, we would slowly help create a healthier environment.

Corinee Saban *is headstrong and passionate about achieving her goals, which at this point include a career in medicine combined with world travel. A native of the Philippines, she enjoys exploring San Francisco, playing basketball, singing in the shower, and the company of friends.*

NOT A DROP TO DRINK

Imagine a world where when you turn on your faucet and not even a drop of water comes out. Or simply turning your shower on and not being able to feel the warm water rushing out. Or not being able to appreciate green, lush grass in our local parks. What if our future holds all of these?

Unfortunately, we are not far away from that future. The four-year drought in California is getting worse by the month. Even though the rainfall of winter 2016 has seemed to be nonstop, it doesn't take away the fact that the drought is here. We are far, far away from fully overcoming this problem.

California has had multiple droughts throughout its history and some are known to last a very long time, some even 180 years. This just shows how careless we can be about this natural resource, that the same thing has happened over and over again. It seems like we haven't learned our lesson, but maybe it's not entirely a careless act. Maybe it's because of a great increase in population. People love the Golden State so much that we have attracted more and more people to migrate here from around the country or from other parts of the world. More people means more demand for water, since it is essential to survival. As our population continues to increase, it has contributed to our lack of water. So this is evidence that we are not only facing the drought problem now, but it is almost certain to continue in the future.

The constant increase of the population in the state of California also increases our water usage. We have reached our carrying capacity—even though we grow a lot of produce here in California, we are starting to lack resources. Carrying capacity is when a species has reached its maximum population that can be supported by its resource.

A big thing that contributes to the drought is the agriculture we all benefit from. We produce a lot of food with our rich California soil. Almonds, grapes, strawberries, avocado, oranges, and walnuts are just a few that are grown here in California. As amazing as it is that we produce so many fruits and vegetables, this also means that our farmers use a lot of water. For example, it takes about 1.1 gallons of water to produce one almond—just ONE! Imagine how many gallons of water are used on just almond farms alone.

So what can we do to help avoid this? As a teenager in California, I'm worried about the future. I have done my fair share to try to conserve as much water as I can. I'm taking shorter showers, watering the plants less, keeping a bucket to collect the drops of water to be used to water plants, and changing habits of how much water I use for drinking or for other activities.

I moved to California just a few years ago from Canada, and as a newcomer to this beautiful state, I didn't know much about the weather patterns. Often when I visited my family here it was summer, so I thought it was just natural to lack rain. They didn't talk much about any water problems. When I moved here, I was hit by the appalling news that California has been in a drought for a couple of years now. Back then I didn't watch the news often or ever hear people talk much about the drought. At first I didn't think much of it. I thought that it would rain sooner or later, and that would be it—problem solved. Once again, I was hit with the realization that a few weeks of rain isn't going to entirely solve this problem. This problem has been building up not only for months but for years, and it'll take years to solve unless Mother Nature unleashes all the water she can. As part of the younger generation, I do worry about what the future holds for California's water.

California Governor Jerry Brown is enforcing a continuation of water restrictions on every county in California. Since June of last year, Californians have managed to cut back twenty-five percent of water use. This shows that many Californians have played their part to help solve the problem we all face. I'm sure a question that floats around peoples' minds is: Will this be enough? It might—but it depends on how long the drought will continue.

But the big question is: are we really ready to do what it takes?

Jose Gonzalez *was born in Leon, Mexico, but was raised in San Francisco's Mission District. He enjoys all the Star Wars installments, but thinks the most recent movie,* Star Wars Episode VII, *is the best yet. Jose works hard at school in order to reach his goal of designing video games.*

GLOBAL WARMING AND CARS AND YOU

Do you know what driving cars can do to us in the future? It's like a chain. Everything connects, such as how driving cars is connected to our ecosystem. An ecosystem is when things are benefitting from each other and affecting each other. San Francisco is an ecosystem that has abiotic and biotic factors. Abiotic factors in the city are cars and biotic factors are the people who live and drive around San Francisco. When these two factors interact, they both affect the ecosystem. In San Francisco, people are using cars too much, which is causing global warming. To help global warming we must drive less and take public transportation more.

The human impact on global warming includes things that we do in daily life. For example, using cars every day burns a lot of gas into the air, and millions of people use cars. According to the U.S. Energy Administration, 136.78 billion gallons of gasoline were being used in cars only in the United States every day in 2014. Every time we use gasoline, it affects the ecosystem more and more by polluting the air and affecting the ecosystem of the city of San Francisco through global warming.

The use of cars is causing global warming, which is affecting our daily supplies of food and water through weather changes. Global warming makes the earth warm up. Every time it warms up it can cause floods and droughts by changing the patterns of the rainfall. According to the U.S. Global Change Research Program, the flow of water will decrease in the West during spring and summer. The water levels will be lower in the late summer. Our impact makes the world hotter and it will cause a change in our water supply.

Another way that global warming affects people is by changing our energy use. We are going to need more electricity based on the demand of people using the cooling system because the earth is getting hotter. Also, the change in temperature is changing the patterns of snowmelt and rainfall, which is affecting our renewable energy sources.

Global warming is also affecting our transportation. The way we are affected is through rising sea levels, which causes flooding and destroys airports, roads, rail lines, and tunnels. Something else affecting our transportation are strong hurricanes, which cause evacuations and the damage of transportation and infrastructure. Because of climate change, people are also getting illnesses and facing near-death situations caused by extreme heat waves. Global warming's effects are causing limited supplies to go away, such as water, energy, and transportation sources. This is like the tragedy of the commons, which is when people have a limited supply and people have to get some of the resources for their daily life, but every time they get supplies, there are fewer supplies for everybody to survive.

People think that driving won't affect anything, but they don't realize how bad global warming can turn out. To help global warming in the ecosystem of San Francisco, people can take public transportation or walk to their destination. By doing this, it will slow global warming down because people do not have to all take out their cars and drive. I've lived in the Mission District for ten years. I walk to wherever I have to go. By walking I am not using a car, not burning gas, and not polluting the air. Since San Francisco is small, I think everybody can walk or take the bus to go where they want to go. If people don't use their cars, they won't be burning gas, which leads to global warming and can cause heat waves. Global warming won't stop, but not using a lot of cars will slow down global warming.

In conclusion, we, the people who live in San Francisco, should use fewer cars so we don't have to deal with global warming and its weather changes, which can affect us in many ways. Some of these ways are near-death experiences caused by heat waves or by the destruction of infrastructure that are caused by hurricanes. Other ways global warming can affect us is by the change in weather, which gives us limited water and also makes us use more energy. In order to stop or at least slow down global warming, we have to walk or take public transportation, and not use cars in a small city. Global warming is happening because of us, and in order to stop it, we need to compromise and work together.

Raymond Jurado *was born in San Francisco, California. He is the youngest of three siblings. He is fourteen years old and a freshman at John O'Connell High School. Raymond enjoys playing basketball, as well as video games. His favorite game is* NBA 2K16. *He really wants to work on his ball control and dribbling skills to improve his game.*

WE MUST CONSERVE!

According to writer Jeffrey Ball's article in *newrepublic.com*, because of the drought in California, "Farmers are ripping out crops, religious believers are praying for rain, and local governments are ordering restaurants to stop serving glasses of water, except to diners who specifically request them." I think it's bad that it hasn't rained enough lately. It affects everybody because we all need water to live. People wasting water are making the drought worse, so they should try to conserve. Throughout this essay, I will be discussing the tragedy of the commons and how people are being selfish by caring only for themselves and wasting water.

In California, the drought has been going on for five years. That is quite a problem. California is a giant ecosystem which depends on its biotic and abiotic factors to work together. Humans are biotic factors, and they use up too much of the abiotic factors (water), and cause the environment to be destroyed. Humans can have both positive and negative impacts on our natural ecosystem.

In dealing with the drought in California, people are the biotic factors that affect the water by wasting it. I noticed that some people don't care that they're wasting water. They daydream or listen to music instead. They must think that there is a lot of water.

I saw this first hand when I was eight years old. My mom asked me to wash the dishes. I left the sink on while I was trying to wash the dishes and the sink flooded, and water was pouring down everywhere. I didn't realize that this was wasting water until I got older. I used to think there was a lot of water and it couldn't be wasted, and now I realize that it's precious.

Humans could change their behavior by not wasting water on fake lakes. According to *cbsnews.com*, homes with boathouses built around artificial lakes have been seen in Indio, California. They wasted lots of water to make a giant lake next to their houses. I think it's ridiculous that some guy thought making a fake lake was a great idea! People are also wasting water by leaving sprinklers on for an hour, which is really unnecessary. It's idiotic to leave a hose on for a long time. The severe drought has impacted rivers, lakes and snow packs. If we continue this behavior, California will be as dry as the Sahara Desert.

I think that the best way to solve this problem is to get everybody on the same page so we can all stop wasting water. I've noticed that when people go to jail, they come out and change their behavior. So maybe if people got threatened with jail time and getting fined, they would start to conserve water. I feel that what we can do as a society is start to make small changes, like putting your alarm on before you shower to help remind you to get out at a certain time. I noticed another thing: while I'm brushing my teeth, I'm being more aware about shutting off the water instead of leaving it on the whole time. This might not seem huge to other people, but I can do my part.

As you can see, I think people need to work together to save water. If I can conserve water, set an example, and inspire my friends and family to do it, hopefully this inspires other people to make a difference by conserving water. If this happens, the drought in California will be less of an issue.

Josue A. Vasquez *recently visited Guatemala, where his dad used to live. He has also visited Mexico, where his mom's side of the family is from. Religion is important to Josue and he is close with his family, including one older brother and one younger brother. They all look out for each other. Right now, Josue is focused on keeping his grades up. He hopes to one day be a lawyer and visit Spain.*

CHANGING THE DARK SIDE OF FACTORIES

Everything that makes our lives easier comes with a price. The world we live in today is filled with inventions. The Industrial Revolution made it easier for us in many ways, such as obtaining the food that we consume, getting from place to place in less time, and having access to clean drinking water. Before, people did not have access to any of these things that we use today. We are responsible for taking care of our ecosystem, like the neighborhood's forest, animals who live here, and also the people who live in these places. By growing plants, raising and killing animals, and polluting the air with factories, we affect this ecosystem. The rise of factories also affected the human ecosystem in a bad way because of child labor. Before inventions, we had to do things ourselves. In my opinion, without technology we would not be as productive as we are today.

Technology runs and changes the world we live in today. There is a mutualistic relationship between technology and consumers; they both need each other. Children working in factories don't benefit from this. There is no kind of mutualism between children and factory owners. It is more like predation. According to the Child Labor Public Education Project, about 2.7 million healthy lives were lost due to child labor, especially in agriculture. Some of the hazards children are exposed to are pesticides, working with machinery, sharp tools, lack of clean water, and beginning to work at an early age. With a mutualistic ecosystem, people help each other so it can become more of a win-win situation. In nature, a meerkat watching out for the other meerkats collecting the food is an example of mutualism. My plan is to create mutualism between the factory owners and the government. What I would do is make sure factory owners only hire adults over eighteen, and have the government help them expand their businesses.

We have things that make our lives faster and easier. But how do we get these things? In the first factories, children aged five to seventeen would work for low wages or for free. The owners controlling the factories would make money out of these children in their own interest, while these children suffered. This is a form of predation.

Today in the world, I've heard that children work on cocoa plantations obtaining cocoa seeds for people who buy the children. If someone I knew got sold into plantations, I would be in shock because the conditions are very risky. Children there use machetes without any protection. They could cut themselves while getting the cocoa seeds. Our lives are easier because we just go to the store and buy chocolate, while children are suffering, and children in the plantations don't even get to taste the cocoa seeds. Plantation owners affect the children's ecosystems, kidnapping or buying them from their families. Children in sweatshops, factories, and plantations may make lives easier for U.S. citizens. Our lives are also easier because of advanced technology. Child labor is what we don't pay attention to. We just consume, consume, and consume.

In conclusion, our lives are easier because of transportation, food, and clean water. We have it easier than the children in sweatshops, plantations, and factories. We just get up, walk to the store, and buy the item we want, which may have been made by child labor. With inventions, our lives are easier, but creating mutualistic relationships between the plantation owners and the government would help the children not be at risk at these plantations. I use technology every day, including my iPad, my Playstation 4, and my laptop. I am grateful for what I have, but maybe some of the materialistic things I have are because of child labor. My goal is making people more aware of what they are consuming and using in their daily lives.

Krissia Martinez

was born in San Salvador and she immigrated to the United States on Halloween of 2012. Actually, the plane ride was a little scary! She loves to draw realistic portraits of people. After graduating high school, Krissia wants to go to Stanford to study to be a general surgeon.

EL SALVADOR, CALIFORNIA, AND SOUTH AFRICA'S DROUGHTS

Imagine for a moment how a drought looks. The soil looks dry, cracked, and grey. The land looks dead, with no trees or plants. You feel thirsty in the back of your mouth. Your skin feels like it is on fire, and inside your head you feel a terrible headache. This is what parts of South Africa look like right now. Now imagine the same place before the drought started. Fifteen years ago, for example, there was a lot of water in South Africa. It was very green, and the plants were beautiful and colorful. There were many kinds of flowers and trees. Animals were drinking water and staying healthy. Animals slept under trees or they made nests in the tree branches. Would you live in a place with no trees and dried dirt, or in a place with a lot of animals, plants, and water? Droughts in South Africa, as well as California and El Salvador, are negatively affecting the people in those areas. But they could get help from other places like Costa Rica and Puerto Rico. If these places communicated about new ideas, these three places could deal with the droughts better.

When I was in third grade, I asked myself about how a drought would look in El Salvador, South Africa, and California. When I was in El Salvador at the age of nine, I was a little innocent girl with a lot of curiosity. When

I started learning about droughts I felt scared, sad, and also I felt like we humans could change our lifestyle to stop wasting water. When I was nine years old, many other people were thinking the same things. Also, people were asking themselves what they could do to stop the drought.

National emergency drought levels were declared by the president in El Salvador. El Salvador has the worst drought conditions in Central America in the last thirty years, and El Niño made the drought worse from 2015 to 2016. People in rural areas were the most affected because they are mainly low-income families, subsistence farmers, and farm laborers. This crisis affects half a million people. The UN says El Salvador's drought is a humanitarian crisis. In El Salvador they are "developing public sector and civil society partnerships to confront vulnerabilities exacerbated by climate change," according to the UN. This means that everybody is sharing ideas to deal with the drought. "Communities developed their own network of emergency response to make their communities more resilient to climate change," the UN says.

In South Africa there was an emergency in six provinces. South Africa is having a strong drought. This affects water supplies across the country; 2.7 million households have trouble with water in their homes. Those who are living on ranches, in forests, and in open spaces are the most affected because they have to collect water.

In California's drought, water levels dropped in local lakes. Governor Jerry Brown introduced mandatory water cuts for the first time in California's history to cut water by twenty-five percent. The past four years were some of the driest on record. The Central Valley, "America's bread basket," is in "exceptional drought status." The Natural Resources Defense Council and the Pacific Institute said, "California could reduce its water use by 17 to 22 percent with more efficient agricultural water use, including fixes like scheduling irrigation when plants need it and expanding drip and sprinkler irrigation."

In conclusion, people in El Salvador, South Africa, and California should take the good ideas the people from Costa Rica and Puerto Rico are sharing with them about how to deal with drought. People in those three places have to be better prepared for a drought because millions of people and animals are dying—they need resources, like water and food to survive. That's sad. People from other places, for example Costa Rica and Puerto Rico, can help them by giving good ideas on how to survive in a drought or help them with resources like food, clothes, and more.

Sandra Rojo

goes by Sandy and is eighteen years old. She was born in Mexico and came to the United States two years ago. Sandy enjoys playing soccer, she is the goalie on the school team at John O'Connell. She has a cat named Kakan that cuddles with her at night. Sandy is considering going to UC Berkeley for college, and hopes to be a dentist when she grows up.

GOODBYE WINTER AND RAIN, HELLO SUMMER

Global warming has become a dangerous problem for our ecosystem. The main cause of global warming is the overuse of energy and fossil fuels by humans. I believe that we should educate people about what global warming is, what is causing global warming, and how we can change our lifestyle to help slow down global warming.

Have you wondered what is happening to our world? Why is our weather getting warmer each year? How people are a threat to the Earth's atmosphere and are causing global warming? Throughout many years, scientists have proven that many biotic factors (defined as living things) and abiotic factors (defined as nonliving things) are creating global warming. NASA defines global warming as "a gradual increase in the overall temperature of the Earth's atmosphere, generally attributed to the greenhouse effect caused by increased levels of carbon dioxide." This is caused by people using too much energy.

I chose to write about global warming because this topic is important for me, and not only for me, but also for all people who live on the planet. Some people are trying to do the best for the planet by learning about global warming and finding solutions for it. Others think only about themselves. This an example of the tragedy of the commons, "a situation where individuals are acting in their own best interest but at the cost of the group's success. Everyone involved is eventually hurt." The solutions will not work if everyone doesn't agree to change their lifestyle.

The first time I heard about global warming was three years ago, when I lived in Mexico. I had a teacher back then whose class I really didn't like, but when he started to talk about global warming I got interested. I felt that I wanted to know more about global warming and what would happen in the future. I remember how Mexico's climate started to change, causing damage to the ecosystem. An ecosystem is defined as a natural system or group of interacting elements formed by the interaction of a community of organisms with their environment. It was getting hotter each year. The time when it should have been raining, it was hot. When I moved back to California I noticed that the climate had changed here, too.

Much of the information known about global warming comes from a study and research by NASA, which proves that global warming is a huge crisis. Some people disagree because some of NASA data show that Earth has experienced climate change in the past without help from humanity. So why interfere now? From tiny samples of the Earth's atmosphere, scientists have built a record of Earth's past climates or paleoclimates. The paleoclimate record reveals that the current climate is changing ten times faster than past global warming has changed climates.

I think the best possible solution to global warming is that people should stop being selfish. People just think for themselves and they are not thinking about the other people that they are hurting. This is an example of the tragedy of the commons. If we stop being selfish we can make changes and make our ecosystem better for us. I think that if people start to turn off their cars for just ten minutes a day instead of idling it would stop many tons of carbon dioxide emissions per year. If a million people recycle, this would probably stop 1,200,00 tons of carbon dioxide emissions per year. Also if people can plant trees in their yards, the trees will absorb 160,000 tons of carbon dioxide emissions per year. These are some few things that we can do to slow down the global warming.

In conclusion, global warming is a really big problem. The earth is something that we shouldn't mess with. Global warming makes plants die, dries out rivers, and makes our ecosystem warmer and warmer. This is a big problem, which will affect us—no water, no food, no animals, no humans. What do people know about global warming? I surveyed my friends and family and this is what they said: Global warming has been happening forever, some answered. Others didn't even know it was happening. Amazing, right?

We can make a change if we recycle properly. We can buy fewer water bottles, use more bikes than cars, and plant more trees, and you should see how our ecosystem will change. We can provide a safer and better life for all. The solution depends on everyone participating.

Damian Hernandez, *a native San Franciscan, lives in the Mission District with his parents and younger brother. He likes playing video games, watching television, and is on the wrestling team at John O'Connell High School. He wants to one day study Japanese and live in Japan.*

RUN! TSUNAMI!

Everybody knows that a tsunami can be dangerous to the coastal ecosystems of countries such as Japan, India, and China. For example, on March 11, 2011, a magnitude 9 earthquake rattled northeastern Japan and caused a devilish tsunami causing hundreds of millions of dollars in damage. Tsunamis are inevitable, and we cannot prevent damage, but there are ways to minimize damages and loss of life in coastal ecosystems.

A tsunami is caused by a series of movements in the ocean floor that can result in an earthquake. Tsunamis can also be caused by a volcanic eruption or—very rarely—a large meteorite strike and an underwater landslide.

Tsunamis can happen anywhere in the world, but "occur most often in the Pacific Ocean and Indonesia because the Pacific Rim bordering the ocean has a large number of active submarine earthquake zones. However, tsunamis have also occurred recently in the Mediterranean Sea region and are expected in the Caribbean Sea as well," according to the NOAA Center for Tsunami Research.

Tsunamis can have devastating effects on coastal ecosystems and their biodiversity. They disrupt the equilibrium of the ecosystem, affecting the biotic (living) factors (people, animals, and plants) and abiotic (nonliving) factors, such as buildings. Tsunamis can destroy human lives and also damage buildings, endangering the environment. They can endanger human life by drowning people and crushing them on impact. Tsunamis can also damage businesses, schools, and homes.

In the Japanese tsunami of 2011, in the village of Rikuzentakata, 1,750 people died. There was also damage to buildings in Japan. The most life-threatening thing was the meltdown at the Fukushima Daiichi nuclear power plant, which was caused when the tsunami flooded the power plant reactor. Because of the nuclear power plant disaster, 100,000 people were evacuated from their homes to ensure their safety, according to the World Nuclear Association.

Most people think that the tsunami damage comes from the first wave, but it actually comes from the flood wave after the tsunami. According to the Sms Tsunami Warning system, "Most of the damage is caused by the huge mass of water behind the initial wave front, as the height of the sea keeps rising fast and floods powerfully into the coastal area. It is the power behind the waves, the endless rushing water that causes devastation and loss of life."

Tsunamis cannot be prevented, but there are ways to prevent deaths of people and damage to buildings. One way for people to stay safe from tsunamis is to have warning signs. According to the NOAA Center for Tsunami Research, "The State of Hawaii is addressing tsunami risk through the Hazard Education and Awareness Tool (HEAT), a Web site template that uses Google Maps technology, spatial hazard data, and preparedness information to help increase awareness of coastal hazard." But while warnings can prevent loss of life, they cannot prevent damage to the equilibrium of the ecosystem.

Along some tropical coasts, forests can be planted to take the impact and stop the damage of tsunamis. According to the Food and Agriculture Organization of the United Nations, "Recently, mangroves, and other types of coastal forests and vegetation have increasingly been reconsidered as possible alternatives to be used instead of, or in conjunction with, artificial structures. Mangrove forests are vegetated inter-tidal wetlands that provide goods and environment services, including protection against wave impact erosion." Mangroves are not an invasive species. They are a native species, and planting them stops the impact of a tsunami, reducing damage while preserving the original ecosystem.

Damage can also be prevented by not building dangerous structures like nuclear power plants along the coast. With all the damage that happened in Japan in 2011, the possible release of radiation from Fukushima would have created the biggest tragedy for the ecosystem. This would have disrupted the equilibrium of the ecosystem, and it could take decades for the ecosystem to recover.

We cannot prevent all damage from a tsunami, but there are ways to prevent some damage to minimize loss of life, loss of human property, and damage to the ecosystem. The best approach would be different depending where you are. In urban areas, think about not building dangerous construction along the coast. In rural areas, think about planting forests along the coast. If we do this, we can preserve native ecosystems.

Kyle Padua *was born and raised in San Francisco, where he lives with his mom and dad in Bernal Heights. Kyle plays basketball and likes to watch TV shows with lots of action, like* Lost *and* The Walking Dead. *He loves to eat pho, and knows where to find the best Vietnamese restaurant in South San Francisco. When he grows up he wants to work as a California Highway Patrol officer to help keep Californians and visitors safe.*

DROUGHT CAUSES CALIFORNIA TO SINK

There is a drought affecting Californians, and it is causing damage to the ecosystem of California. California has been in a severe drought for four to five years and, as of now, state reservoirs have reached a historic low. Governor Jerry Brown has issued a state of emergency, calling for water use cutbacks in all regions of California. One of the reasons for this drought is the gradually rising temperatures in the Northwest. As the drought continues, long-term impacts such as land subsidence, seawater intrusion, and groundwater depletion are being observed. We need to conserve water in the state of California in order for us to have water for future generations.

Did you know that there are many ways in which the drought can impact our lives? Not only that, it can also have a huge impact on the ecosystem that we live in. Our ecosystem consists of everything from plants, to animals, to human beings, air, water, and mineral soil.

The drought has many effects on the ecosystem we live in. For example, plants and animals rely on water to survive, but due to the drought, organisms aren't getting the amount of water they need. Other impacts include migration of wildlife to areas that have water, loss of wetlands, destruction of wildlife habitats, poor soil quality, and lower water levels in ponds, lakes, and reservoirs.

The drought is causing land subsidence in the Central Valley of California, costing billions of dollars. The main cause of this incident is groundwater pumping by farmers. Farmers are using the same amount of water that they were using before the drought for their crops. They really don't look at the big picture. Therefore, this is an example of tragedy of the commons. Over-pumping can lead to thousands of destroyed well casings and canal linings. Sinking a foot every year, this slow motion land subsidence isn't expected to stop any time soon.

There are ways to replenish groundwater in the Central Valley. One way to replenish groundwater is to wait for rainstorms. This could provide excess surface water, which could then be guided from streams via existing water conveyance systems onto dormant agricultural fields. This method is called infiltration, which is the oldest method, but scientists have developed additional ways to replenish groundwater. Direct injection, where you take treated municipal water and inject it to the ground, is another way to replenish groundwater.

As we enter our fifth year of the drought in California, state water regulators are asking people to cut back on water use by twenty-five percent. As water users still use the same amount of water that they are used to consuming, water fines have increased. Fines of $500 have been issued to four communities for consuming excessive amounts of water. Communities such as Beverly Hills, the City of Indio, the City of Redlands, and the Coachella Water Valley District have gathered fines up to $61,000 each for consuming tremendous amounts of water. Beverly Hills plans to add additional fines, and has come up with its own penalties if water conservation efforts do not reach their goals. The interaction in this relationship is another example of tragedy of the commons. wherein human beings take what they need and do not consider the needs of the entire community.

It is important to encourage water conservation in the state of California. This is because we need to look out for future generations. I think that fines for using too much water should rise to make the citizens of California look at the big picture. Raising the fines will gradually decrease the amount of water consumption by residents in California. By doing this, I believe that it will make us stay within our ecosystem's carrying capacity. This happens when species take the resources they need and leave the rest for others. Another way to cut back is to reduce the amount of water consumption. While I think that this solution is helping tremendously, if we reduce the consumption of water a bit more, then there will be a little more water for future generations. Whether we live in the city, suburbs, or rural areas, we all have to do our part and hold ourselves accountable for saving water.

Stephanie Benavides *is a tenth grader at John O'Connell High School. Born in Miami, she has lived around the Bay Area, and currently lives in Oakland. She is an avid reader, loves contemporary dance, and is known for her comical faces that make others laugh. Her ambition is to attend college to study history and law to learn more about the world and help others.*

IMAGINE

The weather has been acting very unnaturally lately. Our ecosystem, a biological community of interacting organisms and their physical environment—our world—is being damaged by our careless actions. We need to do something to stop climate change before it causes irreversible damage. Human behavior has caused climate change, which has resulted in extreme weather events and health problems.

Climate change is a result of people not taking care of the environment and polluting it. Climate change is caused by the increased amount of carbon dioxide (CO_2) in the atmosphere due to the use of fossil fuels such as natural gas, coal, and crude oil. In our ecosystem people, one of the biotic factors (living organisms), are using these nonliving, or abiotic, fuels that are damaging our planet.

It's been found that higher temperatures, flooding, droughts, powerful hurricanes, and more intense tornados have occurred in our world at a higher frequency and out of season due to climate change. For example, this past December in Dallas, Texas, there were several unexpected tornados. These tornadoe created winds that spun up to 200 miles per hour, and destroyed 800 homes. In Texas, tornados often occur in the spring and summertime, but they are very uncommon for December. That's an example of out-of-season weather that has been occurring in the past few years as a result of global warming. These natural disasters are partially caused by pollution because of the factories and cars that are powered by fossil fuels, another example of how our world is being damaged by our actions.

An additional example of global warming in California is the drought. There has been a severe increase in temperature, which has caused the drought—a severe lack of water these past few years. This absence of water has forced the state to go on water restrictions, such as taking shorter showers. If we didn't damage the earth, we wouldn't be in this position. Global warming is caused by humans. We created the California drought and each drought is getting more severe with more serious consequences.

One in every eight deaths can be linked to air pollution. More people are being diagnosed with asthma. In China the air quality is so poor that you can wipe it off the windshield of your car. Breathing that air would be like smoking thirty packs of cigarettes a day. All these issues are caused by the burning of fossil fuels in cars, buses, trains, and factories that burn coal.

I think we should change the way we look at the problem. We need to think more of what we can do to solve the problem. We should reduce the amount of coal we use and replace it with alternative energy, like solar, wind, and hydroelectric energy. I also think we should plant more trees because they give us more oxygen by processing the CO_2 and converting it into oxygen that we can breathe. But the problem is we produce more CO_2 than can be processed. At the rate that we're producing CO_2, we have surpassed our carrying capacity.

Governments really need to work together to come up with laws, agreements, or strategies to solve the climate change problem. Recently, there was a conference in Paris about climate change that I read about in a *New York Times* article. At least 195 countries signed the Paris Agreement last December to pursue laws to limit the temperature increase to no more than 1.5 degrees Celsius. Now, we just need to take action.

I feel I can make a personal difference in stopping climate change by educating my community about trees and nature's role in the environment. For example, there was a really beautiful tree in front of my house and my neighbor cut it down for absolutely no reason. It affected me very negatively because we are already having more than enough problems in the environment and cutting down trees is not going to make it better. They cut it down and didn't think about it twice. They acted as if the tree was personally hurting them. That's just what happened to one tree in one neighborhood. Imagine what it's like in the rainforests where they've been cutting trees for years.

The world we live in is being damaged by climate change because of our actions, which are causing health problems and extreme weather. If we actually commit to helping the environment, imagine how our world would be. It would be greener, the air would be more breathable, people would be healthier, it wouldn't be as hot, and the weather wouldn't be as extreme. Imagine how beautiful the world would be.

JUST LIKE THE BEAUTIFUL, COLORFUL, DIVERSE SEA

An educator's guide to using this book

A Note from Our Partner Teachers
by Vanessa Siino Haack and Francesca Zambrano

The essays in this book came out of the first year of an English and biology integration for ninth and tenth graders. In order to support our students, John O'Connell High School has a fully-integrated co-teaching model featuring the added support of a Student Success Coach in each period, as well as a Special Education co-teacher in one period that has a number of students with special needs. Throughout the year we have implemented project-based learning in our classroom by developing authentic and challenging unit projects that require demonstration of both English and biology mastery targets.

This unit focused on ecosystems, the various relationships within ecosystems, and the benefits of biodiversity. Students also looked at negative impacts of humans on ecosystems and worked to develop possible solutions. We opened the unit with simulations that required students to take on the roles of various types of scientists (hydrologist, naturalist, botanist, and zoologist) and analyze sets of data to determine what was going wrong in the ecosystem and to trace the negative impact back to its source.

After writing an in-class essay on this two-week investigation, students chose one of three essay topics that allowed them to more deeply explore ecosystems. Their investigations and research became the essays you see before you. As students researched and began writing their essays, they continued to investigate biological concepts such as biodiversity, ecological relationships, and invasive species. The topics allowed them to either use their knowledge of ecosystems as a metaphor for social functioning or to more deeply probe the effect that humans have on ecosystems and what we can do about it.

Essential Questions

1. How do I affect my environment?

2. How does my environment affect me?

3. How can I use the power I have in my relationships and communities to effect positive change?

Essay Prompts

1. Biodiversity: Using what you know about biodiversity in ecosystems and group behaviors, analyze whether biodiversity is as advantageous in social situations as it is in biological situations. Think about a time when you were in a mixed group of people (age, experience, ethnicity, gender, etc.). What effect did this diversity have on individual success? What threats are there to diversity where you live or go to school? How does this relate to your environment/social community?

2. Social Communities as Ecosystems: Analyze the ways in which your neighborhood or other social community functions like an ecosystem. What roles do different biotic and abiotic factors play in this "ecosystem"? In what ways is equilibrium maintained/disrupted? What effect does this have on individuals or "species"?

3. Human Impact on Ecosystems: Think of one specific way you see human behavior affecting natural ecosystems, either in the world at large or right where you live. How do you think humans should change this behavior, right now and in future generations? What would the world look like if these changes happened?

GLOSSARY OF TERMS

Abiotic factor: a nonliving thing that influences an environment, such as mountains and oxygen

Biodiversity: the variety of plant and animal life in the world or in a particular habitat or ecosystem

Biotic factor: a living thing that affects an ecosystem, such as plants, animals, and parasites

Carrying capacity: the maximum population size of the species that an ecosystem can support depending on its resources

Commensalism: a type of ecological relationship in which one member benefits and the other is neither harmed nor helped

Competition: a type of ecological relationship in which two organisms need the same limited resource; both are harmed

Consumers: organisms of an ecology food chain that receive energy by consuming other organisms

Decomposers: organisms, often a bacterium or fungus, that feed on and break down dead plant or animal matter, thus making organic nutrients available to the ecosystem

Ecosystem: a natural system or a group of interacting elements formed by the interaction of a community of organisms with its environment. Every ecosystem consists of abiotic and biotic factors that make it a complex form

Invasive species: a plant, fungus, or animal species that is not native to a specific location, and which spreads rapidly

Mutualism: a type of ecological relationship in which both members benefit; a win-win relationship

Parasitism: a type of ecological relationship in which one member benefits at the expense of the other

Predation: a type of ecological relationship in which a stronger organism (the predator) feeds on a weaker organism (the prey)

Producers: organisms in an ecosystem that produce biomass from inorganic compounds and get energy from the sun

Rule of ten: only ten percent of energy that is conserved as food is consumed across a trophic level

Tragedy of the commons: a situation where individuals act in own best interest but at the cost of the group's success

Instructional Activities: Ecosystems

Before they started writing, our young authors used these tools and activities.

ECOMUVE AND FIELD GUIDES

Ecomuve is a free software program available to download from the Harvard Graduate School of Education, which has two learning modules available (pond and forest) that allow students to simulate collecting and analyzing data about an ecosystem and determine the cause of a problem that arises. Because the computers we have in class do not support the *Ecomuve* software, we pulled the data into tables that we printed for students. To keep track of their progress through the *Ecomuve* simulation, we created a Field Guide notebook for our students to use in class daily. It was divided into four parts.

PART I Data and Data Analysis: included space each day for both Observations (Data) and Inferences (Analysis).

PART II Glossary and Field Guide: provided space for students to take notes on biology vocabulary/concepts and the organisms they encountered in the simulation.

PART III Field Reports: provided space for students to write a daily one-to-two-paragraph response about their data analysis.

PART IV Final Report: used for their in-class essay. Students worked in groups of four and each member was assigned a scientist's role so that they became an expert in a specific data field and reported to groupmates for group analysis. After group analysis, students were able to revise their initial analysis to take account of the new learning they had done with their groupmates.

CAUSE AND EFFECT GROUP-WORTHY TASK

Once students had collected all their data, we as teachers developed a series of questions that if solved in the correct order, would lead students to the ultimate cause of the ecosystem damage. Each group was given one question at a time and was required to discuss it, come up with an answer, justify their answer, and write it down in their field notebooks before it was checked by a teacher. If the answers were correct and well-justified, students were given the second question, which built upon the first, and so on until they reached the initial source of the ecosystem destruction. We raised the stakes by encouraging competition among groups to increase student buy-in and were rewarded with nearly one hundred percent engagement. Students used the notes they took during this activity to help them write their final essays.

ECOSYSTEM INTERACTIONS CARD GAME

We created a card game to help students understand the different types of ecological relationships (commensalism, predation, parasitism, mutualism, competition, and altruism). The card game itself consisted of both scenario cards, each of which had a brief description and picture of a biological relationship, and definition cards, which included the definition and symbol of each relationship (+/+, +/-, +/0, -/-). In a game that is somewhat a cross of Uno and Speed, students are required to match biological scenarios to definition cards and explain their thinking to their groupmates.

Instructional Activities:
Interacting with This Book

MAKING CONNECTIONS

Choose an essay to read together as a class. As they read, students actively annotate the text with connections to themselves, other texts, and the world. Students are asked to consider the following questions: What does this essay remind you of? Can you relate to the author? What images come to mind as you read? What do you already know about this topic? Where did you learn about it? Does the text present new or different information? What local and global issues are raised? Start with modeling and guided practice of this strategy, and then transition to independent work. Student responses can then be synthesized in a discussion or through writing.

SOCRATIC DIALOGUE

Read an essay together as a class and ask students to take a stand in response to the piece. Students then identify textual evidence that supports their view to prepare for the discussion. The dialogue can occur between a pair of students, as a fishbowl, or you can split the class to present and defend opposing views. The teacher or students can act as facilitators. Establish norms for participation. Consider the following open-ended questions to start, or create your own that are specific to the text: What is the central idea or argument of the essay? What are the causes and effects at play?

How could the chain of events have played out differently? What can be done now? What are other points of view?

ACT IT OUT

Students get into groups of four, choose a social community (family, school, neighborhood, etc.), and write a scene designed to illustrate the relationships between different members and how it functions like an ecosystem in nature. Students should represent at least four of the following biological relationships: commensalism, predation, parasitism, mutualism, competition, and altruism.

CREATE A PUBLIC SERVICE ANNOUNCEMENT

Begin by showing students examples of public service announcements and discuss persuasive techniques. Students then choose an issue affecting their community or world. This could be a topic appearing in the book (such as gentrification or climate change), or another topic they are passionate about. Then students write a script for a public service announcement about an issue affecting their community or the world. Their script should contain a clear purpose, reasons, and relevant facts.

FORCED CHOICE

Place a sign in each corner of the room: strongly disagree, somewhat disagree, somewhat agree, strongly agree. Read the following statements aloud and direct students to move to the corner representing their opinion about the statement.

· "If only we could interview a whale, we could learn a lot" (p. 112).

· "The grumpy cat meme is a very stable member of the Internet ecosystem" (p. 31).

· "I believe that if people cared about global warming as much as the Super Bowl, climate change wouldn't be as huge a problem as it is now" (p. 145).

· "I grew up here and I plan on living here the rest of my life" (p. 155).

- "Diversity is important to a sports team because it can lead to success, create leaders, and teach social skills to players" (p. 53).
- "People with money and power have always done things without considering or caring about the people they affect" (p. 80).

Call on students with different views to explain their stance. Students will then select a statement that they had a strong reaction to (positive or negative) and read the essay it came from. As an extension, students can write a note to the author. What is your reaction? Why? What is or isn't effective about their essay? What questions do you have for the author?

Content Standards

This project-based unit was designed to address a broad array of standards in English Language Arts and Science, including the following.

WRITING

CCSS.ELA-LITERACY.W.9-10.1 Write arguments to support claims in an analysis of substantive topics or texts, using valid reasoning and relevant and sufficient evidence.

CCSS.ELA-LITERACY.W.9-10.2 Write informative/explanatory texts to examine and convey complex ideas, concepts, and information clearly and accurately through the effective selection, organization, and analysis of content.

CCSS.ELA-LITERACY.W.9-10.4 Produce clear and coherent writing in which the development, organization, and style are appropriate to task, purpose, and audience.

CCSS.ELA-LITERACY.W.9-10.5 Develop and strengthen writing as needed by planning, revising, editing, rewriting, or trying a new approach, focusing on addressing what is most significant for a specific purpose and audience.

CCSS.ELA-LITERACY.W.9-10.6 Use technology, including the Internet, to produce, publish, and update individual or shared writing products, taking advantage of technology's capacity to link to other information and to display information flexibly and dynamically.

CCSS.ELA-LITERACY.W.9-10.8 Gather relevant information from multiple authoritative print and digital sources, using advanced searches effectively; assess the usefulness of each source in answering the research question; integrate information into the text selectively to maintain the flow of ideas, avoiding plagiarism and following a standard format for citation.

CCSS.ELA-LITERACY.W.9-10.9 Draw evidence from literary or informational texts to support analysis, reflection, and research.

SCIENCE

HS-LS2-6 Evaluate the claims, evidence, and reasoning that the complex interactions in ecosystems maintain relatively consistent numbers and types of organisms in stable conditions, but changing conditions may result in a new ecosystem.

HS-LS2-7 Design, evaluate, and refine a solution for reducing the impacts of human activities on the environment and biodiversity.

HS-LS2-8 Evaluate the evidence for the role of group behavior on individual and species' chances to survive and reproduce.

HS-LS4-6 Create or revise a simulation to test a solution to mitigate adverse impacts of human activity on biodiversity.

More Books and Resources for Educators

from 826 Valencia

Check out these previous Young Authors' Book Project publications and other resources, all of which make great teaching tools or gifts for young writers and are available for sale on our website and in bookstores nationwide.

If the World Only Knew: What Fifty-Five Young Authors Believe (2015) contains reflections from ninth graders at Mission High School on their beliefs and where they came from—the people who imparted them, the times when they are most necessary, and the ways in which the world has tested them. It speaks to the power of personal conviction, and why young peoples' voices should be both heard and believed.

Uncharted Places: An Atlas of Being Here (2014) is a collection of stories about place: searching for one's place in the world, places of origin, and places of comfort. Be they metaphorical or explicit, these young authors from Thurgood Marshall High School write about spaces and locales that are important to them, and how we are shaped by place.

The Enter Question: How to Ask and How to Answer (2013) is a collection of essays about, in the words of these authors from San Francisco International High School, "What it is like to start over in a new place." Theirs is a high school for students who have recently immigrated to the United

States, and these essays cover topics including the challenges of communicating in a new language, the courage it takes to ask for help, and the joy of meeting people from all over the world.

Arrive, Breathe, and Be Still (2012) is a collection of monologues and plays exploring the themes of resistance and resilience, written by thirty-five students at Downtown High School in San Francisco. After a semester of working with actors at American Conservatory Theater and writing tutors from 826 Valencia, the students produced this powerful look into the pressures surrounding young people and the strength it takes to keep going.

Beyond Stolen Flames, Forbidden Fruit, and Telephone Booths (2011) is a collection of essays and short stories, written by fifty-three juniors and seniors at June Jordan School for Equity, in which young writers explore the role of myth in our world today. The result is a collection with a powerful message about the stories that have shaped students' perspectives and the world they know.

We the Dreamers (2010) is a collection of essays by fifty-one juniors at John O'Connell High School reflecting on what the American Dream means to them. The students recount stories about family, home, immigration, hardship, and the hopes of their generation—as well as those of the generation that raised them.

Show of Hands (2009) is a collection of stories and essays written by fifty-four juniors and seniors at Mission High School. The authors reflect on one of humanity's most revered guides for moral behavior: the Golden Rule. Whether speaking about global issues, street violence, or the way to behave among friends and family, the voices of these young essayists are brilliant, thoughtful, and urgent.

Seeing Through the Fog (2008) is a guidebook written by seniors from Gateway High School that explores San Francisco from tourist, local, and personal perspectives. Both whimsical and factually accurate, the pieces in this collection take the reader to the places that teenagers know best.

Exactly (2007) is a hardbound book of colorful stories for children ages nine to eleven. This collection of fifty-six narratives by students at Raoul Wallenberg Traditional High School is illustrated by forty-three professional artists, and passes on lessons that teenagers want the next generation to know.

Home Wasn't Built in a Day (2006) is a collection of short stories based on family myths and legends by students at Galileo Academy of Science and Technology. With a foreword by actor and comedian Robin Williams, the book comes alive through powerful student voices that explore just what it is that makes a house a home.

I Might Get Somewhere: Oral Histories of Immigration and Migration (2005) exhibits an array of student-recorded oral narratives about moving to San Francisco from other parts of the United States and all over the world. All these narratives shed light on the problems and pleasures of finding one's life in new surroundings. (Out of print)

Waiting to Be Heard: Youth Speak Out About Inheriting a Violent World (2004) addresses violence and peace on a personal, local, and global scale. Written by thirty-nine students at Thurgood Marshall Academic High School, the book combines essays, fiction, poetry, and experimental writing to create a passionate collection of student expression. (Out of print)

Talking Back: What Students Know About Teaching (2003) is a book that delivers the voices of the class of 2004 from Leadership High School. Previously a required-reading textbook at San Francisco State University and

Mills College, the book helps us all understand the relationships students want with their teachers, how students view classroom life, and how the world affects students. (Out of print)

Don't Forget to Write (2005) contains fifty-four of the best lesson plans used in workshops taught at 826 Valencia, 826NYC, and 826LA, giving away all of our secrets for making writing fun. Each lesson plan was written by its original workshop teacher, including Jonathan Ames, Aimee Bender, Dave Eggers, Erika Lopez, Julie Orringer, Jon Scieszka, Sarah Vowell, and many others. If you are a parent or a teacher, this book is meant to make your life easier, as it contains enthralling and effective ideas to get kids writing. It can also be used as a resource for writers of any age.

STEM to Story (2015) inspires learning through fun, engaging, and meaningful lesson plans for grades 5–8 that fuse hands-on discovery in science, technology, engineering, and math (STEM) with creative writing. The workshop activities are the innovative result of a partnership between 826 National and Time Warner Cable's Connect a Million Minds, and are aligned to Common Core standards.

ACKNOWLEDGEMENTS

A Thank You Letter from the Editors

The Young Authors' Book Project would not be possible without the generosity and dedication of an incredible number of people.

We'd first like to thank the school community at John O'Connell High School for being such welcoming and innovative collaborators on this project. Special, resounding, standing-ovation-style thanks goes to our partner teachers, Vanessa Siino Haack and Francesca Zambrano. These two inspiring educators are doing outstanding work in their integrated English and biology classes, helping students draw connections between literature, writing, and STEM subjects in a way that deepens their understanding of the world around them. Thank you, Vanessa and Francesca, for reminding us all that both the language arts and the sciences are, at their core, about connections and telling stories.

This project would not have been possible without an incredible cohort of volunteer tutors, who spent two sometimes very early mornings a week working one-on-one with the young authors collected here as they wrote and edited their essays. The dedication, passion, and talent they brought to the task was a daily inspiration. Our deepest thanks to: Dana Belott, Jessica Bender, Jennifer Braun, Elaina Bruna, Tehan Carey, Carmen Caserta, Edmund Cavagnaro, Darryl Forman, Kaiya Gordon, Caroline Kangas, Renee Kaufman, Hannah Kingsley-Ma, Marcus Lund, Dina Martin, Kiley McLaughlin, Chiara

Packard, Emma Peoples, Ryan Perry, William Poole, Conan Putnam, Will Randick, Virginia Reinert, Julia Ruiz, Louise Shultz, Maren Smith, Riley Smith, Daniel Spangler, Dave Struthers, Sonja Swift, Amanda Ufheil-Somers, and Kaia Waller.

One group of extra dedicated students and volunteers came to 826 Valencia after school to edit the essays and set the editorial direction for the book. The Editorial Board showed creativity, professionalism, and growth throughout this process, as the project went from being a classroom assignment to an important, lasting publication. Their work shines on these pages. Thanks to these students and volunteers: Elaina Bruna, Tehan Carey, Samantha Gomez, Kaiya Gordon, Gabriela Martinez, Luna Martinez, Izzy Romero-Antoniades, Elijah Romero-Antoniades, Jesus Savaria, Riley Smith, Amanda Ufeil-Somers, Daniel Spangler, and Dave Struthers. Thank you also to Dana Belott, Emily Forbes, Kavitha Lotun, Christina V. Perry, Emma Peoples, Jenesha de Rivera, Ashley Varady, and Jillian Wasick for assisting with editing.

We'd like to extend a deep thanks to Joya Banerjee, our foreword writer for this book. Joya has served on 826 Valencia's board for years and is an incredibly thoughtful, helpful, and wise member of our 826 community. Given her inspiring career in environmental protection and water use and deep commitment to our work, we knew she would do a spectacular job introducing this collection. The way she immediately identified the most powerful, central themes in the students' writing blew us away. Thank you, Joya, for inspiring our young authors in this and their future endeavors, and for making the case for why their work is not only impressive, but important.

Enormous thanks to Tracy Liu, the designer of this book, for being such a thoughtful collaborator and honoring the young authors' words by giving them a beautiful home. To María Inés Montes, our Design Director, and Amy Popovich, our Production Manager, thank you both for your invaluable work on this book, from amplifying the students' voices with your design expertise to keeping us all on deadline. Huge and hearty thanks to Helaine Lasky Schweitzer, our copy editor whose super-human eyes catch every single extra space and misplaced comma, for lending your time to helping our young authors' words shine. Big thanks to Christine Innes for production assistance.

Finally, we are so proud of the students in this book. Young authors, for sharing your unique and poignant perspectives with us, for your courage in offering your stories and voices to the world, and for never giving up on the writing process, we commend and profoundly thank you.

Molly Parent
Programs and Communications Manager

Supporters

We couldn't do this work without all of our donors, including our 2014-15 Shipmates Society leadership supporters, whose generosity makes this annual project possible.

CAPTAINS

Kirsten & Michael Beckwith
Art Berliner & Marian Lever
Daniel Handler & Lisa Brown
Lee & Russ Flynn
Pamela & George Hamel, Jr.
Frances Hellman & Warren Breslau
Mark Lampert & Susan Byrd
Lauder Family Foundation
George C. Lee

Nicholas V. Mori
Michael Moritz & Harriet Heyman
Dave & Gina Pell
Tom Savignano
Michael & Shauna Stark
Andrew Strickman & Michal Ettinger
Laurie & Jeff Ubben
Karen & Jim Wagstaffe

SHIP'S MASTERS

Anonymous
Colleen Quinn Amster & John Amster
David Austin
Joya Banerjee & Harris Cohen
Susie & Sam Britton
Arlene Bucchert & Family
Matt Currie & CM Commercial
   Real Estate Inc.
Gina Falsetto & Warren Brown
Jessica Goldman Foung
Jill Grossman Family Fund
Pamela & David Hornik
Gail & Ian Jardine
Jeri & Jeffrey Johnson

Holly Johnson
Jim & Tricia Lesser
Coltrane & Christopher Lord
Kavitha & Reza Lotun
Frances McDormand
Matt Middlebrook & Lisa Presta
Meridee Moore & Kevin King
Sarah Morrison & Bill Rogers
Derek Schrier & Cecily Horsting
   Cameron
Lee & Perry Smith
Mike Wilkins & Sheila Duignan
Terry Wit
Nicole & Rick Wolfgram

BOATSWAINS

Lawrence & Stacey Bancroft
Jennifer & Nick Bartle
Ethan Beard & Wayee Chu
Josephine Berler
Dominique Bischoff-Brown in honor of
  Different Fur Studios & Patrick Brown
Scott & Jacqueline Botterman
Adriene Bowles
Matthew Bye & Ellen Laguerta Uy
Ninive & Jean Claude Calegari
Emily & Jason Cheng
Chris Clark
Patrick Connolly & Family
Breanna DeGeere
Mark Dempster
Caroline & Warren Dowd
Kelly Dubisar
Isabel Duffy-Pinner & Dickon Pinner
Dave Eggers & Vendela Vida
John Eidinger
Caterina Fake
Girts Folkmanis
Francis-Chapman Charitable Fund
Jon Gans
Michael Glaser & Kristine Hernandez
Green Bicycle Fund
Joe & Barbara Gurkoff Philanthropic
  Fund
Paul Haahr & Susan Karp
John & Marsha Hall in memory of
  Brian Hall
Shepard & Melissa Harris
N & K Hawley
Paul Herman
Liz Hume & Jay Jacobs
Melind John

Diana Kapp
Matthew Kinsella
Jordan Kurland
Michael S. Kwun & Sigrid
  Anderson-Kwun
James & Sarah Manyika
John G. & Lynda W. Marren
Geoff McHugh
Amir Najmi & Linda Woo
Kate O'Sullivan & Kurt Bauer
Dan & Ginger Oros
Barbara Parkyn in memory of
  Anna Picchi
Grant Petersen/Rivendell Bicycle Works
Robin Petravic
Rob Phillips & Elisa Lee
Melissa Powar in honor of Meg Ray
Joan Price
Merle & Leslie Rabine
Michael Rafferty & the Rafferty
  Family Fund
Ramya Raghavan
Matt Rivitz & Blythe Lang
Celia Sack & Omnivore Books on Food
Maté Schissler
Will Scullin
Nicholas Sholley & the Sholley
  Foundation
Staci & Jamie Slaughter
Matthew Sonefeldt in honor of
  Gabe Escovar
Matt Spence
Ken & Kelly Steinthal
Susan Sueiro
Marjorie & Barry Traub

BOATSWAINS (CONT.)

Susan Tunnell in honor of Common
  Sense Media
Ellen Valetta
Valerie Veronin & Robert Porter
Ayelet Waldman & Michael Chabon

Heidi & William Whalen
Sandy & Katie White
Tim Wirth & Anne Stuhldreher
Jesse & Michelle Zeifman

HELMSPEOPLE

PJ & Bruce Bean
Jill & Layne Bourque
Janice Caveliere
Diana Cohn & Craig Merrilees
Kimberly Connor
Amanda Kelso

Daniel Lanza Rivers & Joshua Harris
Derek Linder
Joshua J. Mahoney
Josh McHugh
Steven Miller & Jennifer Durand
Jeffrey Veen

We'd like to extend special thanks to AT&T, the primary fiscal sponsor of the 2016 Young Authors' Book Project. We are able to share these words with you because of this support.

826 VALENCIA

Who we are and what we do

Programs

826 Valencia is a San Francisco-based nonprofit organization dedicated to supporting under-resourced students ages six to eighteen with their writing skills and to helping teachers get their students excited about writing. Our work is based on the understanding that great leaps in learning can be made when trained and caring volunteers work one-on-one with students and that strong writing skills are fundamental to future success.

826 Valencia comprises two writing centers—our flagship location in the Mission District and a new center in the Tenderloin neighborhood—and three satellite classrooms at nearby schools. Both of our centers are fronted by kid-friendly, weird, and whimsical stores, which serve as portals to learning and gateways for the community. All of our programs are offered free of charge. Since we first opened our doors in 2002, thousands of volunteers have dedicated their time to working with tens of thousands of students.

FIELD TRIPS

Classes from public schools around San Francisco visit our writing centers for a morning of high-energy learning about the craft of storytelling. Four days a week, our Field Trips produce bound, illustrated books and professional-quality podcasts, infusing creativity, collaboration, and the arts into students' regular school day.

IN-SCHOOLS PROGRAM

We bring teams of volunteers into high-need schools around the city to support teachers and provide assistance to students as they tackle various writing projects, including newspapers, research papers, oral histories, and more. We have a special presence at Buena Vista Horace Mann K-8, Everett Middle School, and Mission High School, where we staff dedicated Writers' Rooms throughout the school year.

AFTER-SCHOOL TUTORING

During the school year, 826 Valencia's centers are packed five days a week with neighborhood students who come in after school and in the evenings for tutoring in all subject areas, with a special emphasis on creative writing and publishing. During the summer these students participate in our five-week Exploring Words Summer Camp, where we explore science and writing through projects, outings, and activities in a super-fun educational environment.

WORKSHOPS

826 Valencia offers workshops designed to foster creativity and strengthen writing skills in a wide variety of areas, from playwriting to personal essays to starting a 'zine. All workshops, from the playful to the practical, are project-based and are taught by experienced, accomplished professionals. Over the summer, our Young Authors' Workshop provides a two-week intensive writing experience for teenage students.

COLLEGE AND CAREER READINESS

We offer a roster of programs designed to help students get into college and be successful there. Every year we provide six $15,000 scholarships to college-bound seniors, provide one-on-one support to two hundred students via the Great San Francisco Personal Statement Weekend, and partner with ScholarMatch to offer college access workshops to the middle- and high-school students in our tutoring programs. We also offer internships, peer tutoring stipends, and career workshops to our youth leaders.

PUBLISHING

Students in all of 826 Valencia's programs have the ability to explore, experience, and celebrate themselves as writers in part because of our professional-quality publishing. In addition to the book you're holding, 826 Valencia publishes newspapers, magazines, chapbooks, podcasts, and blogs—all written by students.

TEACHER OF THE MONTH

From the beginning, 826 Valencia's goal has been to support teachers. We aim to both provide the classroom support that helps our hardworking teachers meet the needs of all our students and to celebrate their important work. Every month, we receive letters from students, parents, and educators nominating outstanding teachers for our Teacher of the Month award, which comes with a $1,500 honorarium. Know an SFUSD teacher you want to nominate? Guidelines can be found at *826valencia.org*.

People

STAFF

Bita Nazarian
Executive Director

Alyssa Aninag
Volunteer Coordinator

Ricardo Cruz-Chong
Programs Associate

Jorge Eduardo Garcia
Programs Director

Lauren Hall
Grants and Evaluations Director

Allyson Halpern
Development Director

Caroline Kangas
Stores Manager

Amanda Loo
Development Coordinator

Kavitha Lotun
Programs Coordinator

María Inés Montes
Design Director

Molly Parent
Programs and Communications Manager

Christina V. Perry
Programs Director

Emma Peoples
Programs Coordinator

Amy Popovich
Production Manager

Jenesha de Rivera
Finance and Operations Manager

Ashley Varady
Programs Manager

Jillian Wasick
Programs Manager

Byron Weiss
Assistant Store Manager

Olivia White Lopez
Volunteer Engagement Manager

AMERICORPS SUPPORT THROUGH SUMMER 2016

Dana Belott
Programs Associate

Elaina Bruna
Development Associate

Emily Forbes
Programs Associate

Kiley McLaughlin
Volunteer Engagement Associate

Ryan Haas
Programs Associate

BOARD OF DIRECTORS

Eric Abrams
Joya Banerjee
Michael Beckwith
Barbe Bersche
Michelle Yunhi Lee
Jim Lesser
Matt Middlebrook
Dave Pell
Han Phung
Andrew Strickman
Joe Vasquez
Vendela Vida

CO-FOUNDERS

Nínive Calegari
Dave Eggers

It's Always a
Good Time to Give

∞∞∞ WE NEED YOUR HELP ∞∞∞

We could not do what we do without the volunteers who
make our programs possible. It's easy to become a volunteer
and a bunch of fun to actually do it.

Please fill out our online application to
let us know you'd like to lend your time:
826valencia.org/get-involved/volunteer

∞∞∞ OTHER WAYS TO GIVE ∞∞∞

Whether it's loose change or heaps of cash, a donation of
any size will help 826 Valencia continue to offer a variety of
free writing, tutoring, and publishing programs to Bay Area
youth. We would greatly appreciate your financial support.

Please make a donation at:
826valencia.org/get-involved/donate

You can also mail your contribution to:
826 Valencia Street,
San Francisco, CA 94110

Your donation is tax-deductible. What a plus!
THANK YOU!